FLORIDA: A FIRE SURVEY

To the Last Smoke

SERIES BY STEPHEN J. PYNE

Volume 1, *Florida: A Fire Survey*

STEPHEN J. PYNE

FLORIDA

A Fire Survey

THE UNIVERSITY OF ARIZONA PRESS
TUCSON

The University of Arizona Press
www.uapress.arizona.edu

Printed in the United States of America
21 20 19 18 17 16 6 5 4 3 2 1

ISBN-13: 978-0-8165-3272-8 (paper)

Cover designed by Leigh McDonald
Cover photo by Mike Randolph/Tropic Exposure @ Flickr

Library of Congress Cataloging-in-Publication Data
Names: Pyne, Stephen J., 1949– author. | Pyne, Stephen J., 1949– To the last smoke v. 1.
Title: Florida : a fire survey / Stephen J. Pyne.
Description: Tucson : The University of Arizona Press, 2016. | Series: To the last smoke / series by Stephen J. Pyne ; volume 1 | Includes bibliographical references and index.
Identifiers: LCCN 2015035281 | ISBN 9780816532728 (pbk. : alk. paper)
Subjects: LCSH: Wildfires—Florida—History. | Forest fires—Florida—History. | Wildfires—Florida—Prevention and control—History. | Forest fires—Florida—Prevention and control—History.
Classification: LCC SD421.32.F6 P96 2016 | DDC 363.37/909759—dc23 LC record available at http://lccn.loc.gov/2015035281

♾ This paper meets the requirements of ANSI/NISO Z39.48-1992 (Permanence of Paper).

To Sonja,
old flame, eternal flame

CONTENTS

SERIES PREFACE

To the Last Smoke

WHEN I DETERMINED to write the fire history of America in recent times, I conceived the project in two voices. One was the narrative voice of a play-by-play announcer. *Between Two Fires: A Fire History of Contemporary America* would relate what happened, when, where, and to and by whom. Because of its scope it pivoted around ideas and institutions, and its major characters were fires or fire seasons. It viewed the American fire scene from the perspective of a surveillance satellite.

The other voice was that of a color commentator. I called it *To the Last Smoke*, and it would poke around in the pixels and polygons of particular practices, places, and persons. My original belief was that it would assume the form of an anthology of essays and would match the narrative play-by-play in bulk. But that didn't happen. Instead the essays proliferated and started to self-organize by regions.

I began with the major hearths of American fire, where a fire culture gave a distinctive hue to fire practices. That pointed to Florida, California, and the Northern Rockies, and to that oft-overlooked hearth around the Flint Hills of the Great Plains. I added the Southwest because that was the region I knew best. But there were stray essays that needed to be corralled into a volume, and there were all those relevant regions that needed at least token treatment. Some, like the Lake States and Northeast, no longer commanded the national scene as they once had, but their stories were interesting and needed recording, or like the Pacific Northwest or central oak woodlands spoke to the evolution of fire's American century in a new way. I would include as many as possible into a grand suite of short books.

My original title now referred to that suite, not to a single volume, but I kept it because it seemed appropriate and because it resonated with my own relationship to fire. I began my career as a smokechaser on the North Rim of the Grand Canyon in 1967. That was the last year the National Park Service hewed to the 10 a.m. policy, and we rookies were enjoined to stay with every fire until "the last smoke" was out. By the time the series appears, 50 years will have passed since that inaugural summer. I no longer fight fire; I long ago traded in my pulaski for a pencil. But I have continued to engage it with mind and heart, and this unique survey of regional pyrogeography is my way of staying with it to the end.

Funding for the project came from the U.S. Forest Service, the Department of the Interior, and the Joint Fire Science Program. I'm grateful to them all for their support. And of course the University of Arizona Press deserves praise as well as thanks for seeing the resulting texts into print.

PREFACE TO VOLUME 1

IN JANUARY 2011 I CONDUCTED a three-week road tour of Florida. It was an intense immersion into Floridian fire, made possible by the generosity and patience of many old colleagues I visited and of new ones I found. I left knowing far more than when I had arrived, but of course I left with a sharp sense of what I had missed. There are so many sites to visit—so many forms of fire. I could easily have multiplied the places toured and written usefully about each. Like many of those who have passed through flaming Florida, I came away with more fire than I had expected. I thought I might extract three or four essays; I ended with nearly 20 and could readily have doubled that number. I regret particularly the limitations of time and space (I had reached the maximum manuscript length) that prevented me from adding the Disney Wilderness.

Like my other regional reconnaissances, I conceived this volume as an exercise in fire journalism or, as I like to think of them, as history in real time—they are intended, after all, as color commentary. To them I have sought to bring context, particularly a sense of the fire scene as a historical construct. Since these are not academic pieces, I have not tried to impose the same standard of documentation I would for a scholarly text. Rather, I cite sources where I have quoted passages or stated a perhaps counterintuitive fact, identify and thank those people who hosted or otherwise assisted my efforts, and point out perhaps an especially useful work. A note on sources handles the general references. I know only too keenly the number of places left unvisited and the words unsaid. But we write to genres, as the saying goes, and the virtue of the short, essay-driven collection is also its vice.

FLORIDA: A FIRE SURVEY

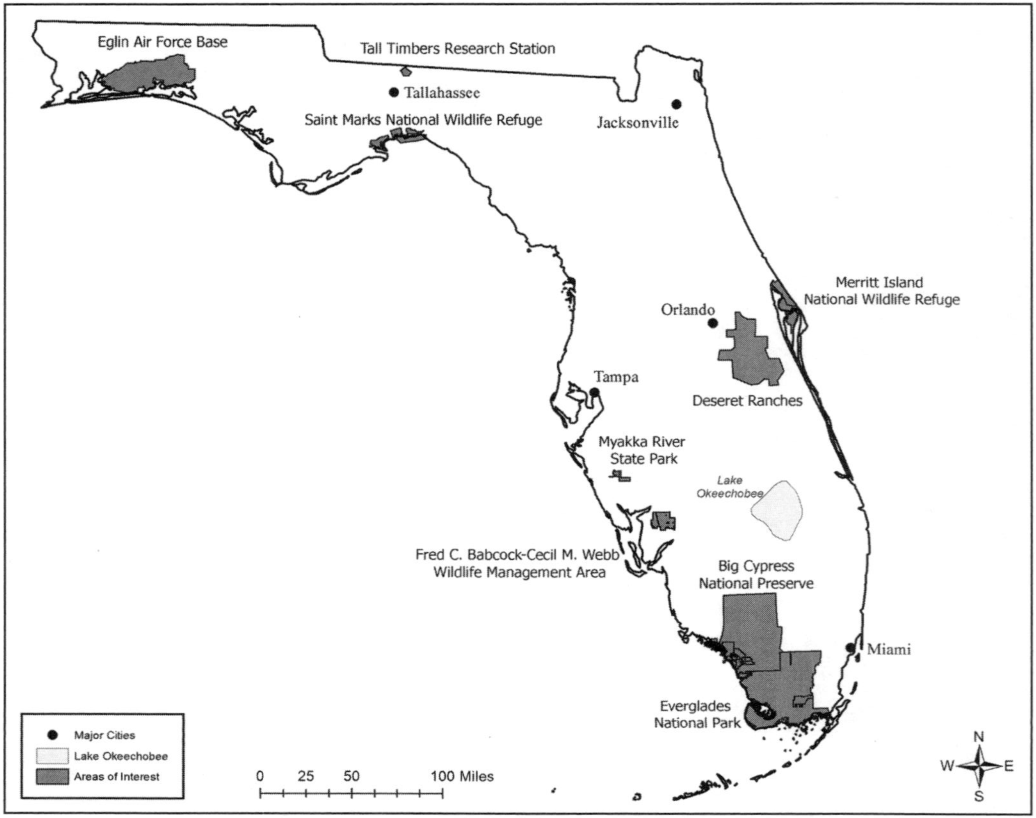

Map of Florida

PROLOGUE

Flaming Florida

IN FLORIDA FIRE SEASON is plural, and it is most often a verb. Something can always burn; something almost always does. Fires burn longleaf, slash, and sand pine. They burn wiregrass, saw grass, and palmetto. They burn turkey oak and live oak, coastal and ridgeland scrub. In wet years the Red Hills and sandhills burn; in dry years, everything does, including hardwood hammocks, big-thicket titi, wetlands redolent with peat, and drained swamps. These are only the indigenous combustibles: the exotic melaleuca, casuarina, and Bahia grass burn just as freely. The lush growth, the dry winters, the widely cast sparks—Florida is built to burn.

There is more lightning in central Florida than anywhere else in North America, and thanks to ever-combustible forage and fast-draining soils, there is more lightning capable of kindling fires. The density of lightning starts spreads out like ripples in a pond; once kindled in the early dry season, before the rains fill up the sumps and swales, or in drought, they can spread for weeks. What nature doesn't ignite, people do: for every fire nature sets, people set five. Ranchers burn to green up pasture. Planters fire sugarcane to speed harvest. Developers clear woods, heap up the slash, and burn the piles. Commercial foresters burn off among their slash-pine row crops. Fire officers on public lands burn off the young rough to keep it groomed and to peel back the accumulated overlays of dense brush and woody debris, a serial firing to nudge the landscape back to health and quell the prospects for conflagrations. Wildlife managers burn marshlands to help waterfowl, longleaf to promote red-cockaded woodpeckers, and sand pine for scrub jays. Hunters burn to encourage the forage and mast favored by turkey, deer, game birds, and bear. Worm grunters burn near sumps to clear patches for harvesting fishing worms. Refuge managers burn to catalyze the habitat needed by endangered panthers, gopher tortoises, and short-leaved rosemary. Not all burns are deliberate, and even when calculated, not all are benevolent; accident and arson add to the overall load. So everywhere the experienced and the cautious seek to shield houses, fragile hammocks, or agricultural fields by burning under controlled conditions combustibles that might otherwise spark a conflagration. For the five millennia or so that Holocene Florida has existed, since the global climate found a mean and sea levels stabilized, Florida has burned routinely, robustly, inevitably, implacably.

Most visitors today identify Florida with its manicured theme parks, golf courses, and beaches. They experience a watery Florida that sweats, pours, and floods, that washes with saltwater from the Atlantic and Gulf, or whose surface glistens with rivers, lakes, and swamps, or is pocked with sinkholes and aquifer springs, or whose air drips with humidity, flashes with thunderstorms, and from time to time seemingly lifts oceans up and spins them inland as hurricanes. But biotic Florida, the Florida whose lush plants inspired its name, is one that burns. It will burn with or without people. The choice is whether the fires will be wild, tame, or feral.

The effort to control Florida's fires is as complex as the effort to channel its waters. The state musters the highest concentration of institutional firepower in America. It has a firefighting capacity equal to that anywhere else, but over the decades it has sought to replace fire fighting with fire lighting, and to an extent rivaled only in a few eccentric landscapes, it has by and large succeeded, although never as much as its fire officers might wish. Every challenge to controlled burning it has met, and often anticipated, and found ways to keep the right kind of fire on the ground, because otherwise fire will fill that vacuum with the wrong kind. In the West endangered species, ranching, air quality, and wilderness cause prescribed fire to falter. In Florida each propels prescribed fire forward.

In some respects Florida is an outlier state, ecologically eccentric and remote in ways similar to Hawaii and Alaska. By other measures, however, it is a bellwether state, an easily accessible place where an influx of people and money has concentrated the promises and problems of American society. That the anomalous and the average have slammed together makes the Florida fire scene as institutionally combustible as its landscape. Fire officers speak less of their aspiration than of their desperation: fire management is not an option, it is a mandate, behind which lies a threat. Unless the land is converted to concrete, it will burn. The only option open to people is to shape how it burns.

Its natural endowment gives Florida prospects to work with fire that few other states enjoy. Its settlement history, which is both very old in time but very young in the pace and magnitude of recent events, has made fire urgent. It has an inherited human presence that bequeathed a rooted culture of fire use and a modern, migratory one that has almost none. If Florida is a leader in American fire management, it is not only because it is blessed with special opportunities but because it has no choice.

Flaming Florida is a dialectic between water and fire.

In one sense the observation is a truism, because fire regimes everywhere involve rhythms of wetting and drying. The wetting grows fuels, the drying readies them to burn. If this were not the case in Florida, fire would not exist there at all. The subtropical summer rains make the land lush with combustibles; the drier winters allow for routine burning as a semipermanent Bermuda high blocks off the daily deluges. This is why so much swampland could be flogged off to absentee buyers. The land was dry part of the year. It only flooded after weeks of rain.

Fires generally follow this annual flooding and draining of the land. The predominantly sandy soils rest on karstified limestone. Surface waters drain or sink quickly; the hydrology is an overlay of streams, lakes, ponds, and seasonal wetlands atop a deep aquifer that runs the length of the peninsula. When the rains are not too relentless, the land drains briskly, soil moisture drops, and burning is easy. But in places this porous surface layer lies atop sheets of impermeable clays, and its thickness varies widely. Where it is shallow, the land will flood quickly; where deep, the land will fill with water more slowly and later. The outcome is a seasonal patchiness of wet and dry scenes that creates a shifting geography of combustion, a landscape of buffers that ensures that some sites will always be available for burning and that fires will only rarely spread everywhere at once. In this way, the surface fuels mimic the seasonal rains, and the burning of organic soil and surface litter helps shape the character of wetlands and water tables.

The place is so wet (an average of 60 inches a year) that something is always growing and available to burn. And equally to the point, it can burn despite wetness that would smother flames elsewhere. Fires can burn through wetlands like marshes, they can combust fine fuels even with relative humidities above 65 percent; duff has even burned at 100 percent relative humidity; and palmetto, always green, will only *not* burn immediately after a heavy rain. A burning index is basically the number of days since rain. A red flag day is one in which the relative humidity drops below 28 percent. Over a calendar year the burning tends to calm at the end of the wet season (say, October), and it flares most ferociously at the end of the dry (say, May). Similarly, it combusts less when the rains fall more and extend past their normal months, and it burns best when drought deepens. But the burning can go on most years day after day, month after month. It's as though the Adirondack forest could burn atop the winter snows or the Mojave Desert after years without rain.

To the rhythms of wetting and drying, absorbed into its biota, natural Florida has added its own peninsular logic. Everywhere in Florida is flat, but nowhere is it far from a major body of water. Land and water heat rapidly, local winds and sea breezes move air continuously, atmospheric mixing is constant. In the winter, for northern Florida, cold fronts blow through. The upshot is that smoke never lingers long. It vanishes into the thick air or lofts away to the Atlantic or Gulf. Rarely does the public, most of which lives along the periphery, experience smoke sufficiently to be annoying. Few know the extent to which Florida burns. Private landowners and public agencies have some decision space denied burners in, say, California, where the San Gabriels and the Sierra Nevada create amphitheaters for public spectacle and basins collect smoke and transmute it into smog.

Over the course of the Pleistocene, peninsular Florida swelled and contracted with the rise and fall of the oceans. Species crowded together or dispersed accordingly. Biologically no less than geographically its tip is closer to the Bahamas and the Antilles than to continental North America; in south Florida the subtropical and temperate meet where the winter frost line wobbles. Around 5,000 to 6,000 years ago the modern climate and sea levels stabilized, bequeathing Florida its natural endowment of flora and fauna. It's worth noting that humans had been in the region for twice that number of years. The place burned, and it has continued to burn without significant pause ever since.

The separate pieces of this pyric puzzle are not unique to Florida, but the way they have assembled to make a distinctive picture is. If you want to understand the ecology of Florida, in the epigram of Ron Myers, you need to know how deep the water and how frequent the fire. The rest, as they say, is history.[1]

Natural history and human history coevolved. Florida is among the most recent of American landscapes, even younger than much of the land that emerged from beneath the Wisconsin glaciation. But it has among the most poorly documented of fire histories since so little natural history is preserved in soil charcoal or fire-scarred ancient trees, no human settlement is carved into stone, and the earliest written documentation is inscribed by fevered conquistadors more ravenous for spoils (or Fountains of Youth) than for natural splendor. The chronicles say little about the region's celebrated timber trees, its now-endangered species, or its endemic fires other than fleeting references to a woodland open and passable.

Florida's natural endowment would guarantee that it burned if people had never appeared. But they did, coexisting with the emerging landscape. That evolved order shattered with contact from Europeans who loosed feral cattle and swine and removed the tending hand of humanity through introduced diseases that evidently decimated the indigenes. The landscape's repopulation by people and their fires played out on a literally unsettled landscape, what Francis Jennings famously described as not a virgin land but a widowed one. By the time European naturalists such as William Bartram toured the region, they were recording scenes that were the still-adjusting progeny of that upheaval. It was a landscape that had evolved to burn. There is no reason to think, however, that its cultural history had not also evolved to burn. Anthropogenic fire is one of the few constants in Florida's human occupation.

The Florida paradox began early. It holds the oldest European settlements; St Augustine was founded in 1565, 40 years before Jamestown and 55 before the *Mayflower* landed at Plymouth Rock. Yet it remained among the most sparsely populated places as successive waves of disease and war hollowed it out and retarded recolonization. The Civil War and Reconstruction further dampened developments. By 1880 Florida was "an outcast among outcasts within the former Confederacy," with a population of 270,000, about a third that of Arkansas. In the post–World War II era, however, it experienced a population influx rivaled only by that in California. The 1940 census tallied 1,897,414 people in Florida. The 1990 census listed nearly 13 million. During the peak decades perhaps 800 people a day were moving in. By 2010 Florida had become the fourth-most-populous state in the country.[2]

All this had implications for Florida's fire history. Almost all the newcomers were urban (or suburban); many were retirees; none proposed to live on the land in traditional ways. Most emigrated from the Northeast and the Ohio Valley—places that lacked a natural basis for fire and had little enthusiasm for continuing burning once the agricultural frontier had passed and industrial alternatives came on the market. The immigrants gathered into urban clusters, primarily along the coast, but no single center evolved into a dominant city-state, thus defying a trend that was typical of the emerging economy nationally. These fire-ignorant (and smoke-allergic) newcomers met a fire-steeped population of residents deeply dependent on burning and resentful of any attempt to regulate their free-ranging fires. This dialectic of newcomer and old-timer worked on the cultural history of fire as much as the dialectic of water and fire did on Florida's natural history.

Each needed the other. Florida has an exceptional history of resisting external introductions but of accepting them when cultivated internally—impermeable to influence from the outside yet porous to ideas loosed from within, even if from transplants. Most forced intrusions were repulsed. Spanish Florida long resisted threats from other European colonizers; British Florida repelled incursions from American rebels; Confederate Florida fought off Union invasions. When the Civil War ended, Tallahassee was the only capital among the Confederate states not occupied by federal troops. Florida more or less successfully resisted Reconstruction. And it might well have persisted as an appendage of the remembered Confederacy. In 1869 panhandle Floridians voted to secede and attach themselves to Alabama, to which it was closer geographically and culturally. (The poll was ignored.) North Floridians generally identified with the Old South, perhaps more after the Lost Cause than before. The state was viciously segregated, at one with its southern cohort. Moreover, until Disney World platted itself outside Orlando, most post–World War II in-migration clung to the coasts, leaving the pith to continue its ways.

Its eccentric geography—the panhandle slicing through Alabama and Georgia, the peninsula plunging toward Cuba—forced a different destiny upon it. The panhandle bonded it to southern folkways; the sunny beaches at Miami and Tampa made it attractive to outsiders, many of whom stayed. Whether they wished it or not, cultural southerners had to make peace with northern immigrants; interior cracker cowboys with suburbanites; the Red Hills with the Gold Coast. And this, in cameo, is what has happened with Florida fire practices. The rooted residents held to their right to burn, which the newcomers learned to accept, and for the most part, failed even to see apart from the occasional rogue smoke across highways. But the newcomers forced those practices to submit to the bureaucratic discipline of permits, the intellectual discipline of science, and the political discipline of an economy that redirected the torch to new purposes.

In no other state have outsiders attempted to impose a fire regime, met with so many failures, and concluded with leading-edge compromises. Each fiat ended with a recantation; each declaration of fire exclusion ended with a program of fire restoration. The chronicle unfolded as though mimicking the rhythm of sunspots.

Behind a loss of control brooded the prospect of massive fires that loomed over the volatile landscape like the threat of hurricanes. In the 1920s large fires washed over the felled pineries. They returned in 1932 and 1934. In 1935 the Big Scrub fire ripped across 35,000 acres of the Ocala National Forest in four hours. The 1950s burned fiercely, with major outbreaks in 1950, 1951, and 1955. In 1956 the Buckhead fire blasted 100,000 acres of the Osceola. In 1971 and 1974 widespread wildfires returned, even though firefighting capacity had improved dramatically. As controlled burning fell behind, big burns hit in 1981, 1985, 1989, and then, making up for a decade's pause, in 1998 they did for the Southeast what the Yellowstone conflagrations did for the Far West. The 2007 fires, widespread but with a focus around Okefenokee, established a new record for big fires in the region.

Against that backdrop ran a succession of failures at outright control that led to a striking litany of national firsts, as Floridians found ways to adapt old practices to new norms. The process began shortly after the advent of state-sponsored forestry. As soon as the U.S. Forest Service tried to instigate a national policy of all-out fire control following the 1910 Big Blowup, the supervisor of the Florida National Forest offered the first coherent internal objection, arguing for a negotiated truce with local woodsburners, who would manage to fire the woods anyway. The federal presence found a local ally in the newly constituted Board of Forestry, or Florida Forest Service; they found common cause in promoting reforestation by excluding fire, and as widespread burning continued through the 1920s, the state became part of the circuit of the Dixie Crusaders, forestry's evangelical branch, with their gospel of fire exclusion. They were ignored. Then they were overruled, not only in the court of public opinion but bureaucratically.

The reaction set in almost as soon as the agency pronounced an official doctrine of fire exclusion. In the early 1930s the St. Marks National Wildlife Refuge became the first in the U.S. Fish and Wildlife Service system to formally accept controlled burning. In 1943 the Florida national forests received the first authorization by the U.S. Forest Service to conduct burning programs. In 1958 Everglades became the first national park to adopt prescribed burning. From 1962 to 1979 the Tall Timbers Research Station, north of Tallahassee, launched its fire ecology conferences, which mustered intellectual arguments for prescribed fire and global counterexamples to the doctrine of fire exclusion; the conferences leveraged a local protest into a national movement. Voluntary associations to promote burning arose, first in south Florida and then in the north; these became the nucleus for an eventual national coalition of prescribed fire councils. In 1977, responding to the proliferation of untended landscapes under absentee ownership, Florida passed the Hawkins Bill, which allowed the Florida Forest Service to conduct prescribed burning for fuel reduction on private lands. In 1990, responding to liability concerns (over smoke as much as flame), Florida enacted a Prescribed Burning Act that, along with a certification program for burners, provided some protection from liability; other southern states soon copied the legislation. In 1999, reacting to the horrific fires of 1998 and possibly crippling air-quality regulations contemplated by the Environmental Protection Agency, the act was modified in ways that expanded control over prescribed fire by the Florida Forest Service and reduced the risks for burners, as it redefined the standard for liability from simple negligence to gross negligence. In 2008, worried over a new rash of challenges, the state's interested parties convened a fire summit to anticipate and answer them.[3]

Over the course of a century, the official context for prescribed fire had gone from legal condemnation to legal encouragement. At each stage Florida had broken trail nationally.

Yet neither its intrinsic disposition to burn nor its stubborn hold on the torch could alone have catapulted Florida into eventual prominence. The catalyst was the newcomers, resented as intruders and coveted as an indispensable source of revenue.

The immigrants brought outside ideas and grafted them onto the indigenous rootstock; Florida was a forced fusion of native and newcomer, and this dynamic applied to fire as much as to politics. Anthropogenic fire would endure, but the cost of survival was to morph in purpose, technique, and ownership. Newcomer notions of social institutions as a value beyond private landownership, of scientific research as a stiffener to folklore, of a broader and more abstract vision that could move beyond provincialism into a national arena—all this happened because an inrushing population forced the issue. The meld was present at the creation of the modern fire regime. The critical Cooperative Quail Study Association, for example, did not emerge from the spontaneous concerns of farmers in the Red Hills, but from northerners who bought hunting plantations and wanted a formal inquiry. Significantly, the association held its first meeting in New York City. So it was Florida, not Alabama and Georgia, that led a campaign for national reform in fire policy.

To an astonishing degree the Florida story tells how outside ideas and money got applied to the state's raw land. The best-known expression is its long real estate boom. An identical, if less fraudulent, process has characterized the stewardship of its inherited natural estate. Native Floridians might identify with the Red Hills, the scrub of the Lake Wales Ridge, the Big Cypress Swamp, and Everglades, but it was Henry Beadel, who had inherited the estate from his wealthy New York uncle, who bequeathed Tall Timbers Plantation as a fire research facility; Marjorie Kinnan Rawlings, from Washington, DC, by way of the University of Wisconsin, who wrote *The Yearling*, a paean to cracker life and landscape; and Marjory Stoneman Douglas, from Minnesota, Massachusetts, and Wesleyan College, who wrote *The Everglades: River of Grass*, which sparked the movement to make the giant marsh into a park. Over and again, major environmental reforms did not well up spontaneously from native Floridians, they emerged from people who had moved to Florida and, perhaps because of childhood experiences, or because they appreciated the sharp contrasts to settings elsewhere, celebrated their adopted landscapes and campaigned for their protection.

And that is also the story of fire. Instinctively, Florida resisted outsider efforts to impose fire policies and practices, but it responded to insiders, old or new. Its desperation for immigration and tourism forced it to accept outside notions, and once naturalized, these became a source of inspiration and a prod to action. But the implacable character of fire in its biota also forced the newcomers to accept burning. Florida's fire mosaic is the outcome of that ongoing dialectic, a uniquely Florida fusion that has become an anchor point in the evolution of America's management of fire on its national estate.

GREATER TALLAHASSEE

HEARTH

American Fire's Silicon Valley

FEW PLACES ANYWHERE have concentrated as much institutional firepower as the Florida Panhandle.

A roster reads like fire's equivalent to the Hollywood Walk of Fame. Among states only California fights more wildfires than the Florida Forest Service (FFS), and none sets so many prescribed fires. The Florida Forest Service has pioneered proactive legislation to reduce hazard fuels, to guarantee landowners a right to burn, and to discipline burning in ways to reduce liability. It is one thing to site-prep the land to receive fire; the FFS has done the bureaucratic site-prep to make fire a treatment of choice. Other Florida state (and county) conservation agencies have aligned accordingly, burning on parks, wildlife sites, and other fragments of the public estate. Among the federal agencies the region is a center of invention and dispersion. The St. Marks National Wildlife Refuge established the first formal prescribed fire program for the U.S. Fish and Wildlife Service, and its practitioners spread their skills throughout the agency. The Panhandle national forests were the first to receive authorization from the U.S. Forest Service to practice controlled burning. Today those forests burn more acres than any other in the national forest system, and they serve as a spawning ground of experienced personnel for the region; the Apalachicola is widely recognized as the gold standard for prescribed fire in the agency. Eglin Air Force Base, formerly the Choctawhatchee National Forest, boasts the largest wildland fire program in the Department of Defense, deliberately burning 100,000 acres of longleaf pine a year, making it a test site for reconciling experimental ordnance with the endangered red-cockaded woodpecker. Even the private sector entwines with the Tallahassee trellis, proving a useful check on the fire establishment. North of Tallahassee, the Tall Timbers Research Station and, farther into Georgia, the Joseph W. Jones Ecological Research Center are the premier private research programs interested in fire. The Tall Timbers Fire Ecology Conferences begun in 1962 are rightly renowned as a point of ignition for America's fire revolution. The Nature Conservancy, which in some years burns more acres than the National Park Service, grew its national fire program from staff housed at Tall Timbers. The idea for prescribed fire councils originated in Florida, and the Coalition of Prescribed Fire Councils has developed its national reach out of the Jones Center. When the U.S. Fish and Wildlife Service inaugurated a national program, it stationed its fire office at Tall Timbers, where it still maintains a liaison. So does the National Park Service; and the U.S. Forest Service houses a regional fire ecologist. Out of this institutional mélange emerged the interagency prescribed fire training manual and certification program; on the southwestern suburbs of Tallahassee, the National Prescribed Fire Training Center sends out 100+ certified burn bosses a year and has trained trainers from 15 countries. The place is an incubator of prescribed-fire startups. Announcing a new collective campaign to promote prescribed fire to the public, the Panhandle collective elected as its slogan "One Message, Many Voices." It might as aptly have said "One Theme, Many Institutions."

Tallahassee sits at one of the three apexes of American fire practices. Specifically, it is the Silicon Valley of prescribed burning.

Of course Florida's environmental matrix is critical. Fire can burn year-round in virtually any condition short of downpours. Combustibles grow like cane, and there is plenty of everything to burn. With some of the highest density of lightning in North America, fire will happen with or without people, or whether they burn astutely or slovenly. Colorado can ignore fire for years except as the occasional disaster. Upstate New York can ignore fire for decades as a relic of settlement history. But fire in Florida is inescapable and in your face. Likewise, other places have a marvelous natural endowment for fire but experience only wildfire. Some other factor has made Florida both prone to burn and frequently burned by people. That factor is the persistence of a peculiar culture of fire.

Or rather, the persistence of multiple cultures that shape-shifted through history but were never extinguished. What they all shared was a folklore of burning, an appreciation that fire was good for them all. There were fire cultures for hunting, for silviculture, for ranching, for land-clearing, for sugar harvesting, for wildlife habitat, for fire protection; for bobwhite quail, Bachman sparrows, scrub jays, and red-cockaded woodpeckers; for cattle, for deer, for wire grass and longleaf pine; for homeowners. They burned to stimulate forage for cattle. They burned to clear out snakes and ticks. They burned for worm grunting. They burned to keep fuel down. They burned to maintain good land and rehabilitate poor land. They burned to protect around structures, from cracker cabins to retirement communities. They burned to remove trash—and trashy landscapes. Riders tossed matches from horseback. Burners in pickup trucks, so the stories go, lit so many matches that they wore off swathes of paint on the driver-side door. Neighbors helped neighbors. Kids learned from parents. Everyone burned because they had always burned. They burned because if they didn't burn their land, someone else would.

It's easy today for the national fire community, keen to promote cultures of burning, to forget the immense social costs of this persistence. A culture of burning endured in Florida because cultural landscapes endured, often in defiance of national economies and norms or even the best interest of local peoples and their land. What the contemporary fire community wants is the fire part abstracted from the rest. It might have happened in other places that kept a folklore of fire, like ranchers along the Flint Hills of Kansas and Oklahoma, or throughout the rural South; but it happened in Florida because here America's northeast and southeast fused into viable if awkward alloys. A persistent tradition wasn't enough: it had to be informed, disciplined, redirected, and capable of broadcasting outside the region. The Florida fusion did that, as practitioners sought to identify, mix, and repackage the best of both regions. In fire this stirring and simmering of ideas, institutions, and practice came to boil in the greater Tallahassee area.

The upshot is a critical mass of fire folks—of researchers, administrators, and practitioners; a synergistic cluster, as industrial policy might call it. They share a common problem on a common landscape, they talk ceaselessly, they transfer not only equipment and ideas but personnel. Staff might start at the Apalachicola National Forest or Florida Forest Service and then join the Nature Conservancy and then go into the Fish and Wildlife Service or the National Park Service and maybe find themselves at Tall Timbers. The fire scene makes a working complex because its members work together. It's the kind of cluster that planners and policy wonks would love to establish elsewhere but cannot decree by fiat. It requires an alchemy of history, landscape, and culture that can't be manufactured. It must grow organically.

This has happened elsewhere. There are other places where the American fire scene comes together with concentrated and catalytic force, where a critical mass leads to a kind of institutional ignition. In such places there is something as it were in the air, water, soil—there is a sense of a fire community, a place where things happen, where the quotidian world of fire practices and the surface routines of fighting and lighting fires seem to leap into the canopy and become the institutional equivalent of a mass fire. Tallahassee is not unique because it has cultivated a complex but for that complex's commitment to prescribed fire.

That's what makes it different from Missoula, Montana, where the U.S. Forest Service houses Region One and the storied Lolo National Forest, along with the Missoula Technology and Development Center, the world-class Missoula Fire Sciences Lab, the Missoula Smokejumper Center, and of course the legacy of the Big Blowup of 1910, arguably the creation story for a national narrative of fire protection. The culture has its folklore, and memorably, it got its poetry when Norman Maclean wrote the 1949 Mann Gulch fire into the national memory. It is a culture of fire in the wilderness.

A similar aura hangs over Southern California. Conflagrations ever loom over the encircling hills, ready to pour over the ridgetops with the next rush of Santa Anas. Against them is arrayed one of the highest concentrations of firefighting forces in the world, and a disproportionate fraction of the Forest Service fire budget. But shoulder to shoulder with the Feds stand the State of California's brigades, CalFire; the mechanical muscle of Orange, Ventura, Los Angeles, and Santa Barbara Counties; municipalities like the City of Los Angeles; and cooperative agreements that can draft engines, crews, and aircraft from throughout the state and beyond. What California's economy is to the global economy, its fire protection is to the global regime of fire protection. Behind that thin red line, the Forest Service has two of its fire research facilities, the Riverside Fire Lab and the San Dimas Equipment Development Center. That the firefights play out in sight of a media mecca means the action becomes instantly nationalized. It is a culture of fire suppression, one that has evolved for over a century, and went on steroids during the postwar housing boom.

Missoula, Los Angeles, Tallahassee—each has, at different times, dominated the national narrative. Today, the Tallahassee model has become the most robust because it is the most innovative, can claim the widest reach, and has a moral fervor almost evangelical in its urgency. Its vital essence is not that everyone burns, although that is what they will tell you and what they do. Its core is that they have learned to rally around a common cause. It is the realization, stiffened into conviction, and then into institutions and law, that whatever other differences they have as land managers, they must cooperate to preserve the right to burn and to ensure that practical opportunities to put fire on the land continue. They know that in the public eye, if one of them fails, they have all failed. Elsewhere, fire often divides. In Florida, it joins.

America's fire revolution bonded fire and land to a common cause, of which prescribed fire became its signatory expression. It was of a piece with other revolutions at that same time and deserves recognition as one of America's genuine contributions to world civilization. What SoHo lofts did for art, what Silicon Valley garages did for electronics, the piney woods of the Florida Panhandle did for fire. Tallahassee was where the action was, where the fire revolution first kindled, and where it has continued to send out its evangelists and entrepreneurs.

AFTER THE REVOLUTION

Tall Timbers Research Station

EVEN AMID CLUSTERS of innovation and times of ferment, some personalities and events stand out: the Robert Rauschenbergs and Willem de Koonings, the Dave Packards and Bill Hewletts, the Steven Jobs and Steve Wozniaks. In the north Florida fire scene, that mantle falls on Tall Timbers, a quail-hunting plantation turned fire research station. The first rush to the barricades, the assault on the Bastille that announced America's fire revolution, began in 1962 when Tall Timbers convened and then published its inaugural Fire Ecology Conference. For the first time critics of forestry's fire exclusion project had a place to gather and voice their shared concerns. Within a dozen years a bureaucratic apparatus that had seemed as monolithic as the Soviet Union, as immutable as the Berlin Wall, had cracked and fallen.[1]

Although Tall Timbers was the first and most visible voice, it was not alone. Among fire exclusion's cohort of critics were wildlife biologists, ranchers and commercial pine planters, prairie restorationists, even the stray government agency, as the Leopold Report published the next year by the National Park Service showed. What Tall Timbers, the place and the conferences, did was give them all a focus: it acted like a giant parabolic antenna, gathering the stray signals and directing them toward a common receiver. It did so in ways that the fire establishment of the day, dominated by the Forest Service and its allies in the state forestry bureaus, could not shut down. It was the brash upstart challenging the reigning monopolists. It outwitted, outmaneuvered, outflanked, outproduced, and outlasted the ancien régime. By 1975 station director Edward V. Komarek could marvel at their success, modestly attributing it to being at the right place at the right time, and concluding that their mission was probably completed.

Margaret Mead famously noted that big changes often came from small groups. That was certainly true for Tall Timbers, as was her observation elsewhere that such revolutions required three ingredients. They needed a charismatic personality, a "sugar daddy" to pay the bills, and disciples to proselytize. Tall Timbers had its patriarch in Edward V. Komarek, a protégé of Herbert Stoddard. It had its sponsor in Henry Beadel, who bequeathed a trust fund and his hunting plantation, Tall Timbers, to the study of fire and its application. And it found ample acolytes among the inchoate collectivity of burners in the Florida Panhandle.[2]

The story of how Tall Timbers Research Station came into being has been told so often that it has achieved the peculiar status of a legend that has become a cliché. The gist is in the late 19th century industrialists, most of them northerners, developed hunting plantations in the south, notably in the Red Hills region of the Florida panhandle and south Georgia. They hunted bobwhite quail in the South as their British counterparts did red grouse in Scotland. But over time the hunting deteriorated. In 1924 the group organized a Cooperative Quail Study Investigation to inquire why, hired Herbert Stoddard, and, again like the British, learned that they had to burn the land. The quail population had thrived in the peculiar ensemble of routinely burned habitats that made up the landscape when the plantations were bought.[3]

Stoddard was the pivot on whom the story turned. Born in Illinois he had grown up in Florida, learned its folkways, and then watched them collapse under clear-cutting, land abandonment, and fire suppression, before returning north for a semblance of a formal education. He embodied a classic American type—the shrewd, self-taught, close-to-the-earth but literate ruralite who perpetuates the best of those origins after the frontier passes. It was Florida's late settlement that made it possible, and the interest of private landowners in someone who understood their setting presented him with the opportunity to give a formal and practical expression to that heritage. What made the story national was that political forestry was moving into the South and proposed a different future. Stoddard published his quail study in 1931; Florida established its Board of Forestry in 1927. The next year the Dixie Crusaders, sponsored by the U.S. Forest Service and the American Forestry Association, traveled the region on the model of peripatetic evangelists, spreading the gospel of fire exclusion to the unbaptized and clearly unrepentant.

The two ambitions clashed. The quail hunters needed a burned landscape; pine-hungry foresters wanted to exclude fire and denounced the tradition of folk "woodsburning" as demented, superstitious, and destructive. Officials wanted nothing that would compromise their fire message. As a later state forester, John Bethea, explained, "If we'd tried excluding fire except sometimes for prescribed fire, the sometimes would be all times with many people." So they argued for the abolition of all burning. Stoddard struggled to get his manuscript published as official editors forced revision after revision, each diluting his conclusion that fire—fire applied more or less in the traditional way—was mandatory for quail habitat (and probably much else that was good in the landscape).[4]

The landowners who sponsored the Cooperative Quail Study were pleased with the results: they burned as Stoddard told them to, and quail flourished. Officialdom, however, condemned the practice and tried to censor the data. Individual plantation owners could not muster an equivalent institutional response, until Henry Beadel converted Tall Timbers from a hunting plantation into a research facility, at which point the burners had a platform from which to reply. Ed Komarek picked up the Stoddard torch; the Cooperative Quail Study went national, and then international, through annual fire ecology conferences; there were even special convocations on Africa and Europe. Resistance crumbled. Like ancient colonizers who carried brands from the mother city to new colonies, Tall Timbers became the hearth from which missionaries carried the torch of prescribed fire. In an eerie inversion, it was as though the Dixie Crusaders had themselves undergone a full-immersion baptism in fire and now preached the gospel of burning to the rest of the country.

During the heyday of Tall Timbers the story of Herbert Stoddard was told and retold until it resembled a ritual incantation, with embellishments here and there like a folktale, because in many respects it had become one. It resonated with the classic formulas of American storytelling. Here was the local boy who held to the best of his rural upbringing, educated himself, and outwitted the professionals. Here was the lone genius, a man of the people, against a hostile establishment overripe with putative experts that ill deserved its status and misused its power. Here was a spokesman for folkways that bested a mendacious modernity bent on clear-cutting the past. The story felt right—it's what many Americans want to hear. The first Tall Timbers fire conference opened with historical reviews by Beadel, Roland Harper, and Ed Komarek that read like the Declaration of Independence's bill of particulars, but the thematic and emotional core belonged to a long memoir by Herbert Stoddard that told of his boyhood education in the burning cracker woods and how the forestry establishment tried to suppress that legacy and failed. Stoddard's saga gave Tall Timbers its creation story and its moral authority.

The Sixties revived the ethos of the rebel, with or without a cause. It was enough to taunt, fluster, and perhaps unhinge the system. The Tall Timbers fire project was never a lark: the stakes were understood as serious by all parties. Yet it fits tidily into a decade when a self-important and overbearing establishment frequently self-destructed through its own malfeasance, lies, arrogance, and hypocrisy.

Most of Tall Timbers's critics were foresters puffed up with their self-identification as professionals and granted political power by government bureaus, exactly the kind of pompous personalities the decade loved to deflate. Official forestry was both dismissive of the early conferences and aghast that decades of public propaganda might be undone. The proceedings of the Tall Timbers fire ecology conferences—not peer reviewed, self-published, freely distributed—seemed to mock the high seriousness of forestry, as though academic ecology might become a *Whole Earth Catalog* and fire science a feature in *Rolling Stone*. The more outraged viewed the whole enterprise askance, as though Ken Kesey's Merry Pranksters had piled out of their bus with drip torches in hand. What would prevent the discipline of prescribed fire, sanctioned only when tightly gripped by official science, from degenerating into the anarchic funkiness of Burning Man?

In fact, Tall Timbers was not interested in mayhem. It had its idealism as befit the times, but it sought to promote a former practice and restore the world which that style of stewardship had sustained, not tear down the present. It gave voice to serious practitioners and scientists. If it succeeded precisely because it was not part of the establishment and could not be co-opted into it, it nonetheless sought to reform the system, not trash it. It lambasted the system to do what the system claimed it was doing—and to allow space for private landowners to do what they thought best on their lands. The vision it offered was a radical restoration.

Interestingly this perspective extended to science, which had also been absorbed by the establishment. Foresters claimed a virtual monopoly over fire science, and the U.S. Forest Service exercised a legal monopoly over federal funding (and had since 1928, the first year of operations for the Florida Forest Service). Either it did the research at its own experimental stations and newly created fire labs, or it sponsored academic research. During the formative years of state-sponsored forestry in the South, it had suppressed unruly fire research with the same gusto it had wildfire. Helpfully, academic historian Ashley Schiff published a damning study, *Fire and Water: Scientific Heresy in the Forest Service*, the same year Tall Timbers hosted its first fire conference. While the book confirmed the insurgents' suspicions, it wasn't necessary: Stoddard and Komarek and their compatriots had experienced for themselves the threat and fact of censorship, and knew how control over funding could indirectly influence not only the direction of research but its character, particularly its preference for abstraction and models.

Just as they countered official propaganda with actual burning on their lands, so the Tall Timbers cohort answered official science with an older brand of naturalism, a radical empiricism in which knowledge came to those who observed closely and honestly for themselves, a form of understanding that derived from living and working on the land for long years. Too often university science was tourist science, coming and going according to an academic calendar; too often government science slid inexorably into political propaganda or political camouflage for bad practices. When participants from Florida State University sought to have the proceedings peer reviewed, Stoddard and Komarek rejected the idea. In their experience peer review was another mechanism by which the establishment squashed protest. They resolved to say what they knew from their personal experience and they were determined to be heard. They edited, published, and distributed the proceedings themselves.

They were not aging adolescent yippies. In 1962 Herb Stoddard was 73 and Ed Komarek 54. They belonged to an older brand of protest, grounded in a stubborn localism, but one to which the Sixties gave a megaphone. They began the conferences a year before President Kennedy was assassinated and decided to end a year after President Nixon resigned. The inaugural conference coincided with Rachel Carson's *Silent Spring* and "The Port Huron Statement" from the Students for a Democratic Society. A year earlier Jane Jacobs published *The Death and Life of Great American Cities*, arguing for America's cities what Tall Timbers did for America's rural lands. A year later Betty Friedan published *The Feminine Mystique*, and the Leopold Report with its alternative vision for the national parks was submitted to the Secretary of the Interior. The *Proceedings* of the first Tall Timbers Fire Ecology Conference belong with those manifestos.

The organizers were themselves astonished at their success. With breathtaking speed, the countercultural became the new normal. The National Park Service revised its fire policies to accommodate fire over the winter from 1967 to 1968; the Forest Service stutter-stepped behind but officially accepted the Tall Timbers mantra that fire management was part of land management in 1974, announcing its acceptance at that year's fire conference, and then brought policy into alignment with the Park Service between 1978 and 1979. When Tall Timbers first began sponsoring a conference a year, its ambition seemed audacious; when the annual events ceased, there were typically several fire-related conferences annually, more than any person could reasonably attend, and where the articles in their original proceedings had made up a sizeable fraction of all fire literature, by the time the conferences wound down, the number of publications had leaped one, then two orders of magnitude, and was rising exponentially.

Somehow, in spite of itself, Tall Timbers had become the national oracle on fire management. Its proceedings served as de facto textbooks. It had forced monolithic bureaucracies to reform, nurtured new start-ups, worked with regional landowners, and even found itself in joint undertakings with the Florida Forest Service. Yet, although the institution kept its cachet, it inevitably aged. Stoddard died, and then Komarek. Tall Timbers seemed to lose its focus: it became a more generalized facility for ecological study, open to researchers of all sorts, and let science determine land use rather than land use science. It got grants from foundations and the National Science Foundation. Peer review became the norm. It lost some of its most experienced land managers. And it lost its nimbleness. When Tall Timbers Research Inc. held its first board of trustees meeting in 1959, the minutes took two pages. Thirty years later the minutes claimed 224. The upstart had become the establishment. In 1989 it was even listed in the National Register of Historic Places.

What happened is the subject of differing interpretations, and as always it is sensitive to whether one compares to what went before or to what came after. But the gist involves a series of institutional missteps or conceptual misreadings or at least an awkward interregnum as the era of revolutionary ardor passed. Each change, reasonable in itself, yielded unexpected consequences when combined with others in what resembled an ecological equivalent to drug interactions as institutional reorganizations expressed themselves as reorganized landscapes. Policy linked management practices such as burning regimens to scientific experiments, not directly to land management. Hardwoods thickened, pine overstories thinned, the biota reordered itself in ways that made the landscape less a "fire-type" by the year, and most shockingly wildlife, its foundational index for good fire practices, was declining. Tall Timbers lost its red-cockaded woodpeckers, was fast shedding its gopher tortoises, and, most ironically, was suffering rapid declines in bobwhite quail. The institution was somehow failing to do what it had long championed as the only true test of fire lore, the ability to make fire work on the land.

If the Tall Timbers story ended here, it could still claim an inextinguishable place in American fire history, and its narrative arc would be no less recognizable. It would have followed that other familiar American folk formula, the generational rise and fall from rags to riches to rags. It would have retold the story of a thousand organizations that had suddenly germinated, found their 15 minutes of fame, and then faded away like old smoke. Instead it did something more remarkable. It returned to the problem of matching fire and land, and by reimagining both fire and land, it renewed itself.

For this it can thank Henry Beadel. Through the terms of the trust and his later will, Tall Timbers had two traits bred into it that made it distinctive. It was a fire research institution, and it was committed to perpetuating a "fire-type" landscape on the estate's 2,800 acres. It both studied fire and used fire to manage what it researched. It was not strictly a laboratory: it was a working landscape that accommodated scientists as it did hunters and birdwatchers. Its management had to express itself ultimately on the land, not in libraries. The sticky issue was that in order to "maintain," it might have to "restore," and in order to restore, it might need to renew. Restoration was easy to envision for the Beadel house; it was less obvious for the Beadel estate. The land had its own logic, and every experiment on it changed that land so that it was never the same landscape twice; science, it would seem, was also subject to adaptive management. The prescriptions that Herbert Stoddard concocted suited a landscape inherited from the post–Civil War era; they did not apply to a landscape set up during the post–World War II era. "Fire-type" meant something different in the 1990s than it did in the 1920s—or the 1960s.

The restoration began in the 1990s. Tall Timbers reorganized, hired an executive director, Lane Green, and began returning to its roots. "Good land stewardship" was reasserted as the informing principle around which all its other activities would cluster. It retired some fire plots (those with long frequencies) that Stoddard himself had established, while reviving the Stoddard spirit, which was less a roster of prescriptions than a close attention to how fire and land interacted. It looked to the land, not to philosophies of land management or theories of fire ecology, to decide what needed to be done, and it determined that fire alone, conducted in the traditional way, was now incapable of resetting the habitat, and so turned to supplemental mechanical and chemical treatments, or more accurately, a prescription cocktail. It hacked the problem trees down, used herbicides, burned hotter fires set in different seasons, and watched the red-cockaded woodpeckers and bobwhite quail return. A working landscape required constant work; maintenance required active intervention. The restoration made fire a catalyst in land management, not its sole determinant.

Something similar happened with the institution's status within the American fire scene. It revived the fire conferences on a biennial and thematic basis. It hosted start-up fire programs for the Fish and Wildlife Service and the Nature Conservancy. It returned as the voice of a tribal elder, not as the voice of opposition. It accepted that it no longer stood alone, that it could no longer influence national policy by protest alone, that it had become interdependent with the enormously complex institutional landscape of American fire, a factious setting that it had itself helped create.

In part its moribund phase was a natural outcome of aging in which the hardwoods of quotidian life crowded in from the edges, and in part it expressed a crisis common to revolutions when their leaders pass away. But part too may have been the result of its own success. In 1960 the forestry establishment was both vast and unified, and as Tall Timbers confronted it, as it rallied and gave voice to forestry's fire critics, its presence grew until it seemed to confront the Forest Service, intellectually, as an equal. From the Red Hills it looked out to the nation, and then to Europe, Africa, southeast Asia, and Latin America for evidence and counterexamples. While it argued its own case for fire, the discourse assumed an adversarial structure, such that the shape of its argument followed the shape of its rival's, not unlike the way Cold War national-security states came to resemble each other. When that opposition collapsed, when it no longer had to lean into the wind, it lost its footing—lost something of its informing identity. In a sense, it lost its way.

It was no longer the little boy saying the emperor had no clothes; the emperor was saying it for himself, and changing clothes furiously by the hour. Critics no longer needed Tall Timbers to give form to their protest. In large measure because of its triumph, there was no single object for them all to press against. The American fire scene had splintered, and then reglued the fragments into institutional particleboard. Tall Timbers had to accept that it was now one among many fire institutions, each of which had its own mission and identity. Tall Timbers was left to discover what its own identity ought to be and accept that it could not be the same institution it was at its origins, any more than the Beadel estate could display in 1990 the same "fire-type" it did in 1920.

Likewise, as Tall Timbers reasserted its role as a fire-type institution, it did so in a very different context. It became a regional leader, doing in the institutional landscape what fire did in ecological ones. It catalyzed. It assisted in campaigning for the Florida Prescribed Burning Act, in birthing prescribed fire councils, in educating the public, in researching the role of fire in carbon cycling, in developing strategic plans for prescribed fire in Florida and Georgia, and in providing an upscale forum for researchers and practitioners, a kind of Davos for prescribed fire. It exercised national influence through its regional presence.

In this way Tall Timbers moved beyond its status as a brash challenger to meld maturity with innovation. It was no longer an upstart, but a nursery for start-ups. It did within the national economy of fire management what IBM, Hewlett-Packard, or Apple had done for consumer electronics. It reinvented itself by shucking diversifying acquisitions and returning to its core competency and founding principles, even when these involved striking shifts from hardware to software. Just what those principles were was something that itself would be subject to a kind of intellectual adaptive management, but its institutional history suggests a likely trend.

If Tall Timbers the land was maintained by anthropogenic fire and the practices that made routine fire possible, Tall Timbers the idea was sustained by a conviction that inspiration, knowledge, and practice came from doing on the land. In an almost Burkean sense, it argued for a deeply conservative revolution, an effort to restore not the details but the temperament of a former time and a former way of knowing, and a conviction that the lived-on land was the ultimate source of wisdom.

INTO THE OPEN AIR

The National Prescribed Fire Training Center

THE NATIONAL PRESCRIBED FIRE TRAINING CENTER (PFTC) is not listed in guidebooks for tourists visiting Tallahassee. There are no billboards urging a stopover, nor highway signs highlighting an exit. It's not on a highway, or for that matter a street: 3250 Capital Circle Southwest occupies the second floor of a retrofitted building off an airport annex.[1]

For someone not in the fire community it is easy to overlook as an institution. Even for members of that community its physical location is obscure. I had a map and drove past it. I pulled over and called for directions. I drove past it again. On the third pass I spotted where I thought it might be. It is not a place that draws you in. It's a place you have to seek out. As an advertisement for a major program—the one national facility dedicated to training in prescribed fire—the structure is a monumental letdown. Architecturally, it's a howler. Its outside says nothing about what goes on inside. Its inside is crowded and the flow awkward amid an aura of high-tech squalor. It's a building designed for another purpose and incompletely remade. In few sites is there such a colossal disconnect between the importance of what goes on and the physical housing that serves it.

Yet few buildings serve their mission so well. Like the larger complex down the tarmac, this is a terminal. It's not a destination but a place for people to pass through. It's a staging area. If it's uncomfortable, it should be. Those who come should want to leave because what the PFTC teaches must be learned in the field. Its students must be eager to be outside its walls. The core lesson it instructs is that prescribed fire is not something studied, it's something done, and something learned by the doing. Prescribed fire happens when people make it happen on the ground.

Probably most people have heard of Parkinson's Law, and most likely identify it with the expression that work expands to fill the time allotted for it. Originally, Parkinson (C. Northcote) evolved his satiric analysis out of observations that the size of the British Admiralty staff bore no relationship to the size of the operational fleet. Between 1914 and 1928 the fleet declined from 62 capital ships to 20 (-68 percent), while dockyard officials increased 40 percent and Admiralty officials 78 percent. An analogous progression applied to the Colonial Office. Hiring appeared to follow some internal law of growth without regard of external factors.[2]

One is tempted to meditate that something similar has occurred within the fire community. The size of the staff has seemingly increased without regard to the number of fires or acres burned. Repeatedly, American fire agencies do not use technological advances to replace people but keep adding machines and people both. While an interesting notion, suitable for scrutiny, this is an inquiry only marginally related to what happens at PFTC. Its true significance lies in another of Parkinson's laws in which he noted an inverse relationship between the productivity of an institution and the perfection of its plans and physical plant.

His examples seem today a bit dated (they were published in 1957). But several are timeless. There is St. Peter's Square: "The great days of the Papacy were over before the perfect setting was even planned." There is the Palace of Nations, built to house the League of Nations, completed in 1937: its design included everything "ingenuity could devise—except, indeed, the League itself" which had "practically ceased to exist." Unmatched among British examples is the story of New Delhi, a new capital that was erected almost in lockstep with British imperial decline. And for Americans there is the Pentagon, not completed until the latter phase of World War II, while "of course, the architecture of the great victory was not constructed here, but in the crowded and untidy Munitions Building on Constitution Avenue." The logic behind these examples is simple:

when institutions are truly busy, they have no time to fret over perfection, whether of plans or physical plants. When they have the attention and money to erect ideal structures, they are on the verge of senescence.

The PFTC can nicely enter Parkinson's ledger on the side of the messy and the overproductive. One wonders where the National Interagency Fire Center, housed conspicuously at the Boise airport and visible to international media, might sit on the scale. It opened in 1969, a year after the National Park Service reformed its fire policy to promote fire by prescription, and its early structures, crafted to stimulate the seamless movement of suppression forces around the country, fully opened just as the Forest Service likewise revised its policies. The nation's Pentagon to fire suppression arrived just as national policies sought to replace fire control with a more varied fire management.

The PFTC is a classic Tallahassee fire start-up. Its version of the company-founded-in-the-garage began when fire managers from the western states were detailed to Florida "to help out with prescribed fire." Joe Ferguson from the Florida National Forests and Pete Kubiak from the Apalachicola District recognized the potential to have such transfers upgrade into a formal training program, rallied like-minded colleagues in the area, and approached the national Forest Service office with a proposal. They got funding for a five-year experiment.[3]

The founding principles were to make the program interagency, to do the training in the field, and to change attitudes—to enhance the confidence and skills to make "every day a burn day." The Florida Forest Service and the Nature Conservancy signed on, the Fish and Wildlife Service agreed to fund a director, the Florida Interagency Coordination Center donated a second-story floor. More cooperators joined; no place was better sited to encourage the fire equivalent of a community barn raising. Moreover, the informal Tallahassee network could muster a large number of public and private landowners willing to provide sites for burning. Each class would spend a couple days in briefing, then head to the field for three weeks of choreographed burning. In keeping with its airport location, it identified hubs of cooperators, reaching as far as Mississippi, Tennessee, and South Carolina. There was always something available. The sessions could run year-round. After its initial trial run, the program received reauthorization.

In its first decade the PFTC racked up impressive statistics. Some 1,170 participants had set 1,623 fires that burned 602,150 acres, over 140,000 acres in the tricky wildland-urban interface. The usual suspects dominated enrollment: 698 from the Forest Service and 352 from Interior Department agencies. But state and local government sent 49, the Department of Defense 23, and private institutions 20. The program had even attracted 28 students from overseas, from Australia to Trinidad, and Sweden to Ghana. After its charter was renewed, it planned to expand operations and specialize some of its offerings. Most importantly, by the testimony of its students, it succeeded in its primary goal. It inspired students to find reasons and opportunities to burn, not excuses not to burn. It instilled burning as the default setting. It crafted 10 take-home messages on that theme, a counterpoint to fire suppression's 10 standard orders. It rose out of boots on the ground, not an architect's blueprint. It projects the Florida lessons learned in very practical ways throughout the country.

In concluding his melancholy survey of the edifice complex, Parkinson doubted it was possible to "prolong the life of a dying institution merely by depriving it of its streamlined headquarters." But he thought it might be possible to prevent an organization from "strangling itself at birth." Examples, he continued, abound "of new institutions coming into existence with a full establishment of deputy directors, consultants and executives; all these coming together in a building specially designed for their purpose. And experience proves that such an institution will die." He cites the idealized building plans for a model organization, the United Nations. "When we see an example of such planning," he concludes, "the experts among us shake their heads sadly, draw a sheet over the corpse, and tiptoe quietly into the open air."

That is not how the PTFC originated. It began in the open air and only comes inside from time to time to catch its breadth. Among its cornucopia of plans, none includes a task-specific, full-service facility. It's too busy.

OUR PAPPIES ARE STILL BURNING THE WOODS

The Southern Culture Behind Prescribed Fire

WITHIN THE SOUTHERN FIRE COMMUNITY there may be a more detested man than John P. Shea and a scholarly paper more loathed than his 1940 potboiler, "Our Pappies Burned the Woods," but I'm hard pressed to know who and what they might be.[1]

In truth, Shea deserved much of that scorn. "Your Psychologist," as he styled himself to his bosses at the U.S. Forest Service, got many of his fire facts wrong. He stood with forestry professionals and dismissed the arguments that southern woodsburners gave him for why they annually fired the landscape. It was the abject irrationality of their views and the mule-headedness with which they held them that inspired the Forest Service to hire Shea in the hopes that he might ferret out the unconscious reasons for their unreasonable behavior. The agency had also approached the American Association for the Advancement of Science about forming a committee of social scientists to complement the inquiry. Margaret Mead and Bronislaw Malinowski were nominally on that board, which sent four participant-observers to live with the natives, to try to explicate their exotic lore, and to describe in proper scientific language how to reeducate them into modernity.[2]

The explanations the local residents gave for woodsburning were all gibberish, Shea insisted, and even if it were possible to find some merits amid the mayhem, those values had gone with the wind and the passing of the agrarian frontier. Its practitioners were now simply another disadvantaged social group, claiming an "uncomfortable place in the class and caste system of the South," and over the past decade had become more flotsam and jetsam washed ashore by the economic storms of the Great Depression. "Ninety-nine percent of them are ill fed, ill housed and ill clothed"—and uneducated. They had no art, no politics, no sense of community, and no future. Family and kin were the only hint of an institutional presence, "the custom of their forefathers" the strongest law. They burned because they had always burned. Woodsburning was nothing more than the "survival of an old culture," now grievously out of sync with the modern world. It gave emotional stimulus previously granted by piney woods pursuits that had lost their value. Woodsburning belonged with folk medicine, squirrel hunting, and snake handling.

Even in April 1940, when Shea published, the ecology of fire in the South was being revisited, and within three years the U.S. Forest Service would allow prescribed fire on its Florida forests. Repeatedly, arguments for fire exclusion had failed when demonstrated on the land, and repeatedly, the authorities, bureaucratic and academic, had insisted that the fault lay with the execution of the experiments, not the theory behind them, like students getting the wrong answers because they failed to follow the directions of a lab exercise. For years the Forest Service shamefully suppressed the awkward results. Even after reforms, the tendency was to work within the forestry guild rather than admit that the locals had known best all along. For the most part the critics played the same game, publishing in scientific journals, participating in professional meetings. Much of what gave the Tall Timbers fire conferences their punch was that they stood outside that tent and said what they thought.

Still, not until he published his memoirs in 1969 did Herbert Stoddard voice his unblinkered opinion that the antifire campaign mustered by foresters and their political organs was "the most intensive—and ludicrous—educational campaign that ever insulted the intelligence of American audiences. It was carried on by well-meaning but utterly misinformed persons." With exquisite irony, Stoddard stood Shea's essay on its head. The intellectuals—professional foresters—were the morons. They had irrationally continued to argue for fire's exclusion simply on the basis of disciplinary traditions, forged in a central European past, that had no meaning in their current time and place. They were the ones prone to superstitious ritual, the ones who received emotional boosts to their humdrum lives from the excitement of firefighting.[3]

Shea could be pompous and self-important, keen to be photographed with a professorial pipe. What made his essay bite, however, was that he dismissed the experiences of a class of people who were poor but had abundant pride; they were in fact the Alabama kin of the Oklahoma dust bowlers. Their descendants were not willing to forget or forgive. Stoddard returned the insult by placing the real ignorance where it belonged and showing that the issue was not over empirical evidence (or "science") but over class and politics. The fire counterrevolution that bubbled up during the 1960s needed an emblem as powerful for its cause as the slinking poor-white woodsburner had been for foresters. They found it in John Shea, who achieved a celebrity in death he never warranted in life.

What is forgotten in the obligatory flogging of his essay, however, is that its core thesis was right. The southern woods burned because it was the scene of a traditional culture of woodsburning. Without the stubborn persistence of that inherited practice, controlled burning would have vanished, as it did throughout the West and on American Indian reservations. If that culture had gone, prescribed fire would likely still struggle today to reassert itself in the Southeast. It would spin its wheels in Red Hills clay just as it has elsewhere.

What made Shea's essay repugnant was not its argument but its attitude, however sanitized with pipe smoke and boilerplate insistence that "democratic" means should be used for reform. A more informed mind and a larger soul could have viewed the scene much differently. In fact, James Agee did, living among almost the identical people, struggling to understand their world on their terms, and publishing *Let Us Now Praise Famous Men* a year later. Agee listened. Shea declaimed. Agee empathized. Shea patronized.

When the truth about landscape burning finally (if grudgingly) came out, foresters initially granted an exception for the southern pineries, and then converted en masse. Quickly, prescribed burning acquired the sanction of science and became an accepted "tool" in the professional forestry shop. In principle it was universal: it did not depend on the quirks of settings and personalities any more than the chemistry of combustion did. Prescribed fire was applied science. It was as portable as a shovel.

But efforts to translate prescribed fire have hit resistances almost as baffling as those that fire exclusion, another nominally universal truth, experienced. Outside the Southeast, where prescribed fire became a foundational strategy, prescribed burning typically struggled. In fact, it thrived only in those places, like the Flint Hills, that had a sustained tradition of burning.

"Blood and soil" is the erstwhile European explanation for the tenacity with which groups identify with a homeland. It refers to long residence at a place but also to the story of their struggle to survive there. Only in parts of the United States does this formula hold; mostly the great American place is the open road and its sacred sites are wilderness parks visited on holidays. But "blood and soil" holds for many American Indian reservations, for Mormon Deseret, for Lone Star Texas, and for Dixie. The Northeast has experienced too many waves of immigration for any group to bond tightly with the land, much less to suffer, endure, or fight over. The West has too much public land, by law uninhabited, which replaces socially grounded blood and soil with an abstract primeval Nature from which people are excluded.

This matters because the South's fire culture is embedded within a cultural landscape, a place of "flame and soil," as it were. It's an organic way of life. Talk about fire with practitioners and they will often tell you how they learned from their fathers or grandfathers. The sons continue to burn the woods as their pappies did. What gets lost in the translation, however, is that they burn in the same places as their forefathers. The continuity lies not merely in the practice but in the place. While contemporary burners may tweak their techniques to do scientific research rather than feed cattle, or to promote red-cockaded woodpeckers rather than turkeys, they are adapting a practice long attuned to a place.

At heart the American fire revolution was about rebonding fire to land. You could not segregate the land from its fires; but neither could you take the fire out of the land. One could abstract woodsburning and launder it through scientific analysis into prescribed burning, but what made it work was not the checklists and written parameters. What made it work was the reconciliation of a cultural practice with a cultural place.

The contrast with the West is striking because west of the 100th meridian public land, not private, dominates, and the population is both recent and transient. The truly incommensurable difference, however, is wilderness.

Wilderness is not the same as pre-Columbian landscapes, although the two were often equated as notions evolved and they continue to merge in the public mind. The lands before European contact were not empty: they were cultural places, and it is an accident of history that diseases and war created the appearance of emptiness to westering Europeans. The New World was older—peopled for a longer time—than much of nominally Old-World Europe. The encountered lands were not fresh and untrammeled, still dewy with the Creation. Moreover, there is ample scholarship to testify that wilderness is more a state of mind than a state of nature. It is an idea that has itself a history. Wilderness is not a land free of culture but a cultural landscape in its own right, which is what makes it so difficult to export outside America, or even outside the extensive public lands of the American West and Alaska.

The assumption that such places have never been shaped by humans has caused believers to overlook or erase the evidence of long residence. It means pretending American Indians did not exist in numbers or did not exercise much power over the places they inhabited, and in extreme cases it has meant physically transporting residents off the site. The received American creation story is one of Old World emigrants confronting a New World wilderness; America's epic is the saga of exploring and pioneering across that wild. Interestingly, the move to preserve "the wild" as a memorial to the American saga is precisely framed by the major immigration statutes of the 20th century. The Forest Service designated the first primitive area (the Gila) the same year the old surge of newcomers was shut down by the Immigration Act of 1924. The Wilderness Act was promulgated in 1964, the year before legislation reopened the floodgates and America received an even larger surge of immigrants.

Such timing suggested that the wild stood outside the usual dramas of blood and soil. In fact, the concept was ideal for a society of immigrants in that no single ethnicity could impose its own story over the national scene. The encountered land belonged equally to everyone or to no one. No tribe of peoples or set of cultural practices could claim it; it somehow transcended them all. The idea sat poorly where people already had long attachments, but the West had been settled differently and more sparsely, and as a national issue everyone anywhere had an equal say. Citizens in New Hampshire and Georgia had as much say over Yellowstone as residents of Mammoth Hot Springs. There would not be—could not be, by definition—any body of indigenous fire practices to preserve along with old-growth lodgepole and grizzly bears.

The only fire that truly belonged was the fire set by nature, a genuinely "wild" fire. Such fires could not be transported elsewhere easily, nor could fire practices like prescribed burning be introduced and find an easy residence. Those places demanded different fires because they occupied different cultural settings that went beyond the simple distinctions between wild and working landscapes. Precisely because they are rooted to the land, fire practices, it seems, are not universal techniques. Flame and soil also have their heritage.

The South was a blooded land, but also a deeply tainted one, not readily transplanted beyond the borders of the old Confederacy, nor was it one that others wanted to replicate. "This land, this South," as William Faulkner put it, carried a heritage of guilt as well as pride. In *The Bear* Faulkner could portray Ike McCaslin being bloodied by Sam Fathers after he killed his first deer, a handing down of tradition, but that initiation was a specific bonding of people and place, and it brought with it the sin of the fathers, which was the selling or other alienation of that land from its people. It was not an experience that would be traded like discount coupons or carried like a pulaski from the Bitterroot Mountains to the Lake Wales Ridge. The South's was a cultural landscape that many observers found peculiar, and not a few even within the region found repugnant largely due to its intractable ordering of society by racial caste.

Its techniques would not migrate well, any more than wilderness has found much traction outside the United States. They could not be plopped into new holes like bare-root saplings. They would have to undergo a long adaptation, a literal recultivation. They would work if they were inextricably bonded to a fire culture, and a fire culture required a fire-cultivated landscape. The deeper the past, the trickier the task of relocating it. Its heritage of fire and soil allowed the South to resist fire-exclusion policies imposed from outside, much as it has resisted the dismantling of its caste system. That same tenacity, however, complicates any ambition to select just those techniques that seem useful from a southern model that works because it is part of an organic whole.

The wonder is not that prescribed fire as practiced in the South has not taken root more widely. The marvel is that practitioners have found ways to spread it at all. Mostly they have reached back to recovered traditions among American Indians or ranchers, or in a few cases have tried to fabricate one out of abstractions. Even among the exceptional South, Florida is distinctive because it has received so much immigration and that flood has splashed over lands not bloodied or remembered. Where the southern tradition has carried beyond the borders of the Southeast it is mostly by way of a Florida conduit.

America has a pluralism of fire cultures. The South focuses on prescribed fire because that is the tradition that survived the long agrarian tenancy of a rooted people. Fathers taught their sons to burn. For woodsburning to thrive in the New New South it had to acquire some discipline in conception and execution, which came through scientific attention, bureaucratic permitting, and formal training. The skill still passes across generations, although now it is as likely to come from agency mentors as from fathers.

The mountain West's fire culture veers toward confronting fire in wildland. It also has its lore and its poetry. There is, for example, that magical passage in Norman Maclean's *Young Men and Fire* in which he portrays the doomed smokejumpers strolling toward the Mann Gulch fire, their heads wooly with daydreams. ("The answer" as to what was in their heads "has to be 'Not much.'") Yet he imagines them as playing the mental game in which they see themselves from the outside, as in a stadium. "Always" among the spectators is their father, "who fought fires in his time." Mostly, Maclean is speaking metaphorically, but not wholly, for the West has a fire tradition as real and vigorous as that in the South, and it also goes from generation to generation.

Its cultures are indispensable to the American fire community. They are what actually light and fight fires on the ground. Yet official training and policy all begin with the pronouncement that fire management is science based, that it derives from truths not bound by places and peoples, that fire knowledge, like crews and air tankers, are subject to total mobility. What you learn about the fire ecology of palmetto can apply to ponderosa; the dynamics of flame through southern rough obeys the same principles as that through Southern California chaparral. It is science that must identify suitable policies and yoke fire cultures to achieve them. It is science that has redeemed fire from the medieval ages of superstition and suppression.

There is little in America's fire history to support such notions. The advent of a "science-based" forestry extinguished indigenous fire lore everywhere it went and has yet to replace that loss with effective practices or to suggest what values society ought to pursue in its management of fire. Instead it has flooded the field with information and created tools and decision charts to process that data. It insists that only it can transcend local fire culture, as wilderness transcends cultured landscapes. Yet like wilderness, science, too, is a cultural creation, and instead of shaping culture, it is shaped by it.

In the end, this may be John Shea's real failure. He claimed, as a psychologist, to stand outside culture. This, after all, is the promise of science, that it can objectively analyze and announce treatments based on evidence not valenced to personality or place. It was a conception that in his day merged seamlessly with an administrative order of command and control. Science informed, management applied. No longer would the whims of history or the eccentricities of local lore compromise a rational reconstruction of the landscape. The new knowledge would transcend generations of trial-and-error experience coded into practice handed from father to son.

That formula has destroyed fire culture (and burning) in landscape after landscape, and will continue to do so where authorities insist we must get the science right before we burn. "The science," however, is always a work in progress: we will never get it ultimately right, if only because the nature we study is not an untouched Nature but a landscape shaped by people, among whom are the scientists themselves. What we are studying is our own past. Science, too, is a cultural invention, and like Shea it must stand within the panorama of what it studies.

The self-identified voices of moderation will rightly plead that science must provide a rational order to the inherited folk culture of burning. Yet we could argue just as forcefully that the practice of burning must shape fire science because it determines what gets studied and how the reductionist data of science returns to the land through practice. Part of the appeal of wilderness as an arena for fire science is that it appears to pare away that cultural component and let science speak to nature unmediated. It appears to offer an alternative fire lore—one not based on any particular people or history—for a place that deliberately excludes accumulated human experience, when the reality is that we have another kind of cultural practice interacting with another kind of cultural landscape.

It is no accident that the most successful prescribed fire programs are those that begin with a culture of burning, or that in wilderness the only fire culture allowed is that of scientists, who can claim to be at the place and yet not be there, like characters out of *Star Trek* hiding behind holographic screens from which they can observe and not interact. Both are fictions. Better to imagine the enterprise as a special kind of cultural creation.

The Shea Syndrome was not unique to the American South. It has repeated itself around the world, from Cape Colony to Krasnoyarsk. It has affected California as much as Georgia. Sometimes it actively destroyed indigenous knowledge, the intellectual equivalent of timber companies clear-cutting through the global forest. Often it sought, with all good intentions, to rebuild landscapes after wars, plagues, and land-clearing had caused them to collapse, like modernist architects designing cities from first principles after conflagrations, bombs, or urban renewal projects had razed them. The rational fire-free landscape has as little utility (or popular attraction) as LeCorbusier's Radiant City.

Because Shea found it hard to put himself into the scene—could not transcend his training—what he thought was convincing ended up as condescending. He misanalyzed the choice as one between science or sentiment—between reason and emotion, or modernity and tradition. In reality the contrast was between learned abstract thinking and empirical knowledge grounded in a place. In dismissing local lore, as John Shea did in Alabama, fire science denied itself a font of data, and fire management misplaced the most practical tool in its shed. Just as it is easier to take fire out of a landscape than to put it back, so it is with fire lore.

The American fire community may find itself in the paradoxical position of having to invent tradition, as the Scottish nationalists did in the 18th century for the Highlands. The process of recovered tradition may prove as bogus as the psychological hocus-pocus of recovered memory. Authenticity, however, has little to do with it. We have come to think of land and people as bonding through fire. But it is equally true that fire and land must bond through people. Whether ersatz or erstwhile, that means a culture of fire.[4]

NOT EVEN PAST

Ichauway Plantation and the Joseph W. Jones Ecological Research Center

The past is never dead. It's not even past.

—WILLIAM FAULKNER

IN 1939 SOME 300 ACRES of former farmland in Baker County, Georgia, was converted to slash pine, the latest of the serial plantations that have characterized so much of southern economies. The fast-growing conifer, laid out like the row crops before it, sat amid a classic old-field rough of grasses, forbs, weeds, and hardwoods, kept at bay by chemicals and mowing. The routine burning that had played over the landscape for centuries was banished since the pines would remain vulnerable to surface fire for 20 years. Instead the plot went through a prescribed cycle of silvicultural treatments, primarily thinnings, before fire rejoined the mix in the 1960s as a low-cost technique for keeping down fuels.[1]

All this was not unusual for the time and place. The Great Depression had crushed marginal farming, and a market for southern-pine pulp had emerged. The southeast was fast replacing worn-out cotton fields with its new commodity crop, pine. What was different was that this particular patch of the land belonged within the complex known as Ichauway Plantation, created as a 29,000-acre private preserve for quail hunting. By the time those slash pines reached maturity, however, Ichauway was itself fast morphing, not unlike the lands it had converted from tenant farms to tree farm. It was now under the direction of the Joseph W. Jones Ecological Research Center (JERC), and the Jones Center wanted the land to return to its roots as a biota defined by wire grass and longleaf pine.[2]

The usual scenario would have been, as in the past, to cut, plow, and plant over. Fell the slash pine and hardwoods. Harrow and herbicide the understory. Sow native grasses and longleaf. Treat the inherited landscape as urban renewal practices of the time did cityscapes. Wipe out the unwanted past and strike out immediately to the future. Ichauway, however, saw another path diverging in those woods. It elected to build on the imperfect, inherited present. It would step, not leap, into the future. Ichauway would evolve incrementally by keeping most of the mature slash and gradually replacing them with planted longleaf. For this the restorers needed fire, and for fire, ample fuels. That made the restoration of wire grass, a lush fuel intimately associated with native longleaf, a critical consideration. The legacy of plow cultivation, however, had stripped the site of wire grass; it would have to be reintroduced through sowing; and to flourish it would need to be burned.

Wire grass was not only itself an important fuel: it platted out a biotic matrix for the others, particularly conifer needles. In the distant past this dynamic had happened naturally as falling longleaf needles got snared in wire grass's fountain of fronds. It would take decades before the reintroduced longleaf could mature sufficiently to serve that role, but the mature slash pine, no longer harmed by surface burns, were already doing it almost as well. Besides, scarifying the site by clearing and planting would only further sour the soil of wire grass and thicken the woods with weeds. Paradoxically, shock treatment would prolong the process of conversion.

Building on the inherited structure, however, would allow wire grass and needle cast to create a synergy that could coat the surface with the combustibles needed to reinvigorate something like the old regimen of fire. Regular, even annual, burning could then wash harmlessly over longleaf seedlings, decimate slash regeneration, and rejuvenate wire grass. Bit by bit, fire would allow longleaf to supplant slash in a virtuous cycle in which the old order contributed the fuels for its own slow immolation. Paradoxically, by persisting, the past would make the future possible.

This new conversion would require patience and a very long view. It had taken 60 years for the slash to mature, and it might well take another 60 for longleaf to complete the cycle. Generations of human overseers would come and go in the meantime. When measured against the cadence of annual or biennial burns, the tenure of their passage might seem long; when measured against the deep rhythm of the longleaf forest, it might seem fleeting; but restoration was a project that no one generation would hold wholly. Its narrative would span beyond the grasp of its creators. In all this the plot might stand as a cameo for the saga of Ichauway itself.

Ichauway Plantation was the creation of Robert W. Woodruff, long-reigning president of the Coca-Cola Company (1923–54) and subsequently a 30-year member of its board. From 1929 until his death in 1985, Woodruff acquired small farms which he transformed into a quail-hunting plantation. At his death, his munificent estate went into the Robert W. Woodruff Foundation, one of whose charges was to maintain Ichauway in something like its existing character. Extensive consultations followed; there was the example of the Tall Timbers Research Station at Thomasville; and the site had been used in the past by Emory University as a field station, most spectacularly for malaria research under Eugene Odum, a founder of ecosystem ecology. The outcome was the Joseph W. Jones Ecological Research Center, established in 1991, as a means to satisfy this wish. It falls within the pale of the Tallahassee hearth.

The legacy of "Mr. Woodruff" is everywhere visible. It begins with the geographic arc of the plantation's lands, reflecting the pattern of his purchases and their landed structures. Rudely U-shaped, Ichauway is a pastiche and palimpsest of past land usage, of old fields, seed plots, hammocks, pastures, wetlands, woodlots, and piney wood savannas. The legacy extends equally to the built environment of houses, barns, roads, and workshops. These would be managed in much the same way as the land. And, not least, the story of the Woodruff legacy follows itself a narrative arc—an archetype—remarkably similar to that of the slash pine plantation.

Woodruff's modest residence remains exactly as it was the day he died. The Ichauway store endures more or less intact, still selling dry goods to the residents of the plantation. Most other edifices have been moved, removed, or rebuilt. At its height Ichauway held some 200 tenant structures; these are now down to 20, rehabilitated and relocated to a central historic zone, and occupied by the plantation's new residents—research scientists, graduate students, and general laborers. In brief, those structures that are no longer necessary have been culled out and the new recreated in its place, all in a process of selective substitution. The tempo of change from hunting plantation to ecological research center is faster than that of the forest, but the two processes share a common model.

It's not hard to see why: this is an anthropogenic landscape. Remove its people and Ichauway would soon overgrow with rough and hardwoods, burn savagely here and there, and change its character. It is the land's ancient occupation by people that has made it something people value: they have shaped it not only to their own interests but in their own image. When the land changes, it must evolve out of the legacy bequeathed to it. Since the process occurs incrementally, the shape and temperament of that inheritance—the past—persist.

Ichauway arose out of small farms and woods. The mix of patchy croplands, open woods, and brushy rough offered an ideal habitat for northern bobwhite quail. Since Reconstruction hunting plantations catering to the wealthy had taken root in the region, particularly around the Red Hills, serving America's elites as the grouse-hunting estates of Scotland did Britain. In both cases the landscape had flourished amid the decline of former agricultural sites. Then decay became decadence, and the population of preferred birds of prey plummeted. Much as Scottish lairds turned to a Grouse Commission, so southern landowners, with the help of the U.S. Biological Survey, created a Cooperative Quail Study to investigate why. The conclusion in both places was the same. The habitat had collapsed, and it could only be resuscitated by burning. Both inquiries concluded that it was almost impossible to burn too much.[3]

So Ichauway burned. Not all of it, since not all was committed to quail habitat. Not equally, since parts were added at different times and came with prior histories that had to be massaged variously into the system. But the prime hunting grounds were likely burned annually between late February and May, after the hunt and before nesting. Set fires were light and backing, and often kindled at night. Other sites burned as possible, some deliberately, some accidentally, some occasionally through lightning. The outcome was an elaborate patchwork of fire and fallow that suited the bobwhite quail and its hunters equally. Fire was a bond between them: both were fire creatures, a fire bird and a fire hunter.

With the advent of the Jones Center the push began to render more of the land into the character typical of its earlier days when longleaf savanna blanketed the coastal plains as tallgrass prairie had the Great Plains. In the late 19th and early 20th centuries, industrial logging and turpentining had stripped longleaf, the fabled southern yellow pine, from the scene. Before then, it had comprised some 60–90 million acres throughout the southeast; now less than 3 percent remained, all scattered in fragments throughout the region. Ichauway itself held only patches; most of the lands it acquired had already been scalped.[4]

It soon became apparent that no single technique could transcend all the varied sites. Most students of the longleaf identify four prime natural locales in which longleaf is found; Ichauway holds three of them. More to the point, the sites have to interact with histories of land use, of which there are many more. One suite of practices is necessary to convert from old fields; another, from slash and loblolly plantations; yet another, from hardwood hammocks. Where longleaf and wire grass still thrive, a quick recovery is possible and can then turn into a point of friendly infection for further spread.

The formulas for regeneration, that is, have to adapt to a much-checkered past. They have to find ways to integrate lands variously plowed, grazed, felled, burned, rooted by hogs, and hunted. Today's overseers would have to build an ecological preserve out of a patchy inheritance as their predecessors had cobbled together a hunting plantation out of small fields and felled forests. They have to fashion working, connected landscapes out of a generous but unspecific bequest grounded in cultural perceptions and ambitions that must reconcile with natural capacities. Erecting laboratories and dormitories for visiting scientists is simple; the baffling task is to express those visions through the medium of longleaf, wire grass, broom sedge, live oak, fire ants, slash pine, soils plowed and eroded, dabbles of wetlands and riparian strips.

This time another firebird serves as an index species. The red-cockaded woodpecker (RCW) is federally listed as an endangered species. The RCW nests only in cavities excavated out of old, living longleaf—a habitat that depends inextricably on fire. Again, in the familiar Ichauway scenario, one set of practices and groups is gradually replacing another. Sown seed plots for quail are giving way to artificial cavities for woodpeckers. Hunters are yielding to birdwatchers. In place of mule-drawn wagons, observers drive SUVs. In place of the daily bag of quail, success is measured in cavities occupied by red-cockaded woodpecker. When the program began in 1999, Ichauway had one male RCW. By 2009 it had 23 active clusters, had approached the state for formal designation as a mitigation site, and was contemplating exporting birds in the future.

Today, both endeavors persist. The hunt for quail continues, on a smaller scale, with English pointers and refurbished mule-drawn wagons. About a third of Ichauway remains committed to quail, and some 30 days a year, to their hunting. At the same time, the hunt for red-cockaded woodpeckers flourishes. Woodpecker habitat is replacing quail habitat much as longleaf is substituting, bole by bole, for slash. Between those two firebirds spans the narrative arc of Ichauway's experience in applied ecology.

The dynamic grout in the Ichauway mosaic is fire. Everything that happens happens with or around fire: reconstructed landscapes must, in fact, be designed so that they can burn, for fire is the primary ecological catalyst and accelerant. The land is not simply a fireshed, but a fire habitat. Almost every purpose for the land requires that it burn, and the only way to get the right regimen for burning is for people to do it—to "put the fire out," as the old-timers used to say.[5]

What matters most is the frequency and continuity of the burning. Not one founding burn, but many over time; not boutique burns, minutely calibrated to season and fuel, but the simple (if brutal) presence of repeated fire—this is what animates and integrates the landscape. Of course the pattern matters; the prescribed fire program at Ichauway recognizes 13 burn objectives and a like number of "habitat-fuel" types. Wire grass will germinate best if burned in the growing season, and longleaf may be killed if its new meristemal growth (colloquially known as "candles") are exposed to flame; but both thrive amid any and all other burns. They will accommodate almost any fire better than they can survive fire's absence.

If this outcome seems counterintuitive, it may be because we take a too-narrow view of fire. We see it as a mechanical perturbation, as part of a plexus of causes and effects that jar, push, volatilize, or otherwise rearrange hydrocarbons (fuels) in ways that force a biota to adapt. Fire is an agent of change—of course. But it is also a means of synthesis. It is as much a product of its setting as a producer of it. It does at Ichauway what it does everywhere: it interacts with its surroundings in ways both catalytic and cathartic. As it propagates, it assumes a fluid morphology by integrating terrain, fuels, winds, humidity, all the elements of its various fire triangles, into a zone of combustion that shape-shifts instantaneously as it moves through a landscape. Whatever is in that landscape enters into the flickering history of its passage. At Ichauway it is history itself that gets integrated.

The means by which this happens should surprise no one: the torch has been handed from generation to generation. Fire practices have replaced one another exactly as have flora and fauna. It began when Robert Woodruff hired Herbert Stoddard, the dean of quail and longleaf management, to advise him how to establish Ichauway as a hunting plantation, and Stoddard passed along the lore he had absorbed as a child in central Florida, and then rekindled during his experience as lead researcher on the Cooperative Quail Study Investigation. Stoddard gave both rigor and credibility to traditional burning techniques. That the region has become the epicenter for landscape burning is due in no small way to Stoddard and his successors at the Tall Timbers Research Center, which initially served as a loose model for the Jones Center.[6]

A Stoddard protégé, Leon Neel, then advised Ichauway on both fire and forestry, and resisted the kind of abrupt conversions that would turn old farms into short-rotation slash plantations. As the JERC was getting traction, it hired a covey of staff from Tall Timbers to help establish its fire program. They carried with them that inherited lore even as they modernized operations. ATVs and driptorches replaced mules and matches; geographic information system mapping supplemented experience acquired with specific sites over long decades; computer modeling assisted transferred knowledge. But it is striking that the pith of the program came not by downloading published scientific data and conclusions from science to Ichauway but by relocating experienced personnel. Its fire staff had grown up in the region, they had learned from master burners, they embodied the art as well as the craft of burning. Of course they appealed to science for advice and to technology for force enhancement; but the success of the program depended on grounded experience. That's what is too often missing from national projects that seek to abstract critical knowledge—"lessons learned"—and promulgate it virtually. It needs to be learned on the ground.[7]

To be sure, the staff is deeply committed to translating that personal lore into the codes of quantitative science. Over time, one can expect that the inherited knowledge will be replaced, byte by byte, by more abstract formulas and will move from personal memory to server farms. But for now they coexist, with one growing out of the other—Ichauway could not function otherwise. When modern foresters spoke of "putting the fire out," they meant extinguishing the flame. When Ichauway foresters said it, they conveyed an older, vernacular meaning, of keeping fire alive on the land.

Most places have followed a history like that of the region's longleaf: they had local knowledge clear-cut. This had the unintended consequences of destroying not only a visible overstory of old-growth wisdom but a tangled understory of diverse lore and the institutional capacity, the social soil as it were, to reacquire understanding. The old was stripped away, to be replaced as the new got planted in the literature of formal learning. In reality it committed the land to what proved a working future a long time in coming, and one rife with stumbles, glitches, and outright errors. Still, it seemed a seductive path through the woods—modern, quick in its dismissal of inherited baggage, its returns apparently rapid.

This was, however, over and again in one mode after another, the path Ichauway chose not to follow. Its fires would connect as well as change. They would burn across history as well as pine savannas; they would link past and future as well as wire grass and longleaf. At Ichauway fire's past isn't past.

THE FLORIDA FOREST SERVICE

Florida's Fire Fulcrum

LIKE FLORIDA ITSELF the Florida Forest Service began late, struggled to become normal, and then emerged, by the third millennium, as both exceptional and exemplary. In the institutional ecology of fire's management in Florida, the FFS is the keystone species. Its name may change, and frequently does with new gubernatorial administrations, switching from Florida Forest Service to Division of Forestry and back, but the significance of its role remains invariant. The system could not have evolved properly without it, and once established it has remained the indispensable agency.[1]

When authorized in 1927, Florida state forestry resembled a typical entry state for the Clarke-McNary program. It inherited a landscape that, as the expression went, was cut over, grazed over, and burned over. The old forests were shattered, new ones struggled to establish themselves. The Florida Forest Service committed to fire control and reforestation. The first state forester, Harry Lee Baker, came by way of the U.S. Forest Service and the State of Virginia. The Dixie Crusaders roared through to help shill the message of fire exclusion. In 1931 the County Forest Fire Control law authorized matching funds by which the state and counties might pay for fire protection—in effect, an internal Clarke-McNary program for the state.[2]

During the Depression the Florida story resembled the southern states generally. There were losses and gains. Land fell tax delinquent, so some counties withdrew from formal protection, but other lands were acquired by the federal government for Civilian Conservation Corps use, and these became the basis for a network of state forests and parks. Myakka River began the process in 1934; in 1939, the Blackwater River joined through a lease transfer from the Department of Agriculture. In 1932, 1934, and 1935 big fires ripped through what remained of the woods or what had patchily regrown. The arrival of a pulp industry lent an additional shoulder to the effort to replenish the pineries. The state gradually built up its contract counties and its capacity along lines that mirrored the national story in which more land came under protection and suppression improved. By 1950, while fires continued to sweep 35 to 50 percent of unprotected lands, they burned only 2 to 3 percent of protected lands. In 1941 the Florida Forest Service officially accepted controlled burning (the U.S. Forest Service followed, in Florida, two years later), although burning did not emerge as an organized program until 1949. Out of necessity the state was slowly shuffling from rear guard to vanguard.[3]

In the postwar era Florida and the FFS began to boom. The FFS extended fire protection over an additional 5 million acres and beefed up its plows, radios, lookouts, and aircraft. Major fires struck in 1955 and 1956, further boosting interest in mechanical capabilities and strengthening ties with cooperators. By 1958 the FFS extended its aegis over 46 of Florida's 67 counties, which along with the state forests totaled some 16 million of the state's estimated 21 million actual or potential forested acres. The Feds helped with the rural fire-defense program and the excess-equipment program that funneled used military vehicles through the FFS for fire control. By 1967 56 counties were enrolled, and, following a mandate from the state legislature, by 1972 all were.

Amid the mid-decadal drought of the 1950s, one of the most severe on record, the amount of land burned by wildfire waned, even as the amount burned by prescription waxed. Aggressive, modernizing fire control, supplemented by prescribed burning, had rudely matched the returning or replanted woods and their rekindled fires. And the FFS assumed sole responsibility for all open burning, whether for forestry, farming, ranching, or clearing. The agency controlled both fire control and fire use. Where most state forestry agencies could legally only put fires out,

Florida's had a say in how they were set. Almost uniquely it could manage fire on private land.

At this point the Florida story splits into exceptionalism. In 1969, following a new state constitution, the FFS was reorganized and renamed the Division of Forestry within the Department of Agriculture and Consumer Services. But the real mover and shaker was the astonishing transformation of the landscape. The conversion of rural Florida into urban Florida continued unrelentingly, and in a muted echo the state itself began to buy land for nature protection and public use. The Land Acquisition Trust Fund was created in 1963, followed by the Environmentally Endangered Lands program (1972), the Conservation and Recreation Lands program (1979), Save Our Coasts and Save Our Rivers (1981), Preservation 2000 (1991), and Florida Forever (2000). The state lands were assigned to four agencies for management; the largest block fell to the Division of Forestry, reorganized just in time to accommodate the windfall.

The three trends converged to make the FFS a triple threat. It could hit fires, run with prescribed burning, and field-manage state forests. Forestry agencies in other states had fire protection responsibilities, some had lands to oversee, and a few did controlled burning on a regular basis; but none intertwined them as the FFS did into a single agency, much less with a force that made the program a national presence. In particular, much as Florida's land acquisition program anticipated the threat to nature conservation, so FFS foresaw threats to prescribed fire and moved to head them off. Just as urban development began to squeeze rural lands, so it started to strangle open burning. In the 1970s prescribed fire reached an annual high of 3.9 million acres before restrictions caused it to fall to half that amount.

Then a double crisis boiled up. On state forests and parks, land management meant fire management, and fire management meant prescribed burning, so any restriction on fire's use kept the FFS from effectively caring for those forests. On private lands for which it had contractual responsibility for fire protection, FFS could not deal with the rank overgrowth that made fires explosive (and firefighting ineffective) and was unable to mobilize controlled burning on the scale required. During the 1970s the growth of state lands and private suburbs both shot up; the FFS reckoned that more prescribed fire was needed or more wildfire would sweep protected and built landscapes equally. As Jim Brenner explained to Dade County when it considered buying a patch of land, "If you can't burn it, don't buy it." In 1971 and 1974 wildfire burned more acres on protected lands than any time since records began in 1928. So even as national pressures made prescribed burning trickier by the year, the Florida Forest Service moved to ensure it could apply fire more or less freely. An escalation of legislation to promote prescribed fire proceeded in tandem with bills to expand protected land.[4]

These seemingly parallel lines crossed in 1977. The new landowners purchased fire-prone sites as earlier buyers had swampland. They did not understand the threat posed to communities hacked out of rough and scrub that was then allowed to regrow; and many were absentee owners, holding the land for future retirement. The untended land encouraged untended fires. A few breakout fires that threatened communities (and hence the state's migration economy) got the attention of the legislature.

The outcome was the Hawkins Bill of 1977, which allowed the FFS to do the burning that landowners were unable or unwilling to do on their own. In effect, overgrown plats were treated as a public nuisance, not unlike vacant lots in a city. No other forestry agency in any state had anything like such authorization. In a place often prickly about defending private property, it was a remarkable concession to the unremitting pressure fire placed on Florida life. The FFS added thousands of acres to its annual routine of burning. Even so, the area burned continued to sag. More lands were up for burning, and fewer got burned.

The reasons are many, and they heightened during a tumultuous decade. Not least among them was the suddenness of the transformation. New forests and parks needed administration, infrastructure, plans, experienced staff, all of which took time and money, and each year added to a backlog of burning. New responsibilities under the Hawkins Bill meant more work for the existing apparatus. The national environmental legislation over clean water, air, endangered species, and so on that flung out like sparks from a whetstone during the 1970s created uncertainty, which led to pauses, which let marginal lands fall to the wayside of prescribed fire. Meanwhile wildfire returned in force to open, close, and define the middle of the coming decade. Each affected a different region of the state, but Florida overall could not avoid them. In 1985 the fires roared out of the rough and into Palm Coast, compelling evacuations and incinerating

131 houses. That got political attention. The fires were both a warning that said the fire organization was not keeping pace with development and a distraction that pushed emergency services ahead of prescribed burning.

Even as a sense of urgency grew, so did restrictions. The old habits—burning done on a kind of open range—were being fenced in by a new society stringing houses and shopping malls through the countryside and by an encroaching legal environment built on the barbed wire of liability. In 1987, attempting to avoid a lawsuit-driven shutdown, the FFS commenced a program to certify prescribed burners. Then in 1990 the Florida Supreme Court ruled in *Midyette v. Madison* that both landowners and contractors doing the burning for them were liable for damages resulting from escaped fire and smoke that, in this instance, led to a highway fatality.

The punch and pace of threats would have deflated most organizations—did overwhelm almost all states, and even the federal fire agencies scrambled to do what their fire policies admonished. But once again, Florida went beyond merely reacting and sought to promote good burning. A blue-ribbon committee consolidated concerns into a single bill, the Prescribed Burning Act of 1990. The legislation is remarkable on two counts. One, it created a disposition to burn. It identifies burning as a property right, considers it in the public interest if conducted under appropriate rules, and limits liability to "general negligence." The second innovation was to leave to FFS the determination of what the guidelines and suitable rules might be. The Florida Forest Service became the keystone agency for responsible burning; this applied even to federal lands because smoke management was a state task under the Clean Air Act. A landowner could burn on his own but had no legal protection. A landowner who submitted to FFS guidelines for training and authorization had basic protections.

Still, the burning lagged. Unburned preserves became ecologically disheveled; wildfires lurked in the urban rough. In 1998 the reckoning came. The year began with record floods, then flipped into record drought. The first fire broke out on the Apalachicola National Forest on May 25. Two months later 2,300 fires had raged through half a million acres, burned down 300 homes, forced the evacuation of Flagler County, and essentially commandeered the national fire suppression apparatus. Some 10,000 firefighters from 47 states were drawn into Florida, along with almost two-thirds of the national air tanker fleet. It was a larger firefighting force than had descended on Yellowstone a decade earlier. Fire was not simply a matter of ecological stewardship or of purely local concern; it thrust itself into the face of government at the highest levels as an issue of public safety. Governor Lawton Chiles established a multi-organization committee to review the crisis and offer recommendations. Strong aftershocks struck in 1999 and 2001.

The upshot was an overhaul of fire protection. This was a normal response: it's what any political entity would do. What made Florida different was that fire officials sought to direct attention to where they believed the greatest need was—prescribed burning. This was not so much a matter of more money as more freedom to operate. In 1999 the Prescribed Burning law was amended to replace "general negligence" with "gross negligence." In practical terms this created a presumption to burn. Everywhere else fire's suppression was the default setting; in Florida, the default option was to burn; suppression happened when the burning stumbled. But the burning bears no more relationship to historic open-range firing than an office park does to a pine island. Its capacity multiplied again when it acquired responsibility for regulating smoke under the Clean Air Act; it acts as the Environmental Protection Agency's agent, such that a prospective burn requires only one application, not many. "One phone call, one five-minute conversation, one smoke plume projection model run, and if all is well, one permit. That's how Florida burns get done." A lot of burning, but burning that operates under a dense halter of institutional discipline controlled by the FFS.[5]

The burner, the burn, the oversight—the FFS certifies, approves, and sets guidelines. To qualify for protection, a prospective burner becomes certified by completing a training course, submitting a prescription for review, conducting the approved burn, and having the outcome inspected. There are limits set by the time of day, by smoke dispersion, by drought conditions, by capabilities to respond to an escape. Practitioners grumble that Tallahassee does not understand the peculiarities of local settings, that the opportunities for burning continue to shrink, that burning gets harder every year. Yet few are reckless enough to burn without the extraordinary protection afforded by the Prescribed Fire Act. One fatality from a smoke-blinded car wreck could bankrupt them financially, politically, and morally. It would tar everyone in the Florida fire community, and if it compromised prescribed fire, it might well lead to many more lives lost from wildfire.[6]

Requests to burn arrive typically by phone as a day presents opportunities. The FFS sets as an internal standard that it will respond within three minutes. When the day is right, the requests to district offices flood in. In 2010 requests for permits came in rushes to burn pasture, stubble, sugarcane, citrus prunings, pine plantations, habitat for wildlife, and habitat for newcomers. There were fires to renew land and fires to clear it for paving. Some were broadcast burns, some were piles. Altogether, the FFS authorized 77,076 fires to burn 2,647,590 acres and 123,116 piles. It would like to see those numbers double.

No one is wholly happy with the system or with the amount burned. Every thoughtful observer would like fewer restrictions and more fire, and those accustomed to more freewheeling times often blame the imposition of the system for the lengthening lag in untreated land. Even the FFS experiences the fire gap: it is one thing to buy land, another to manage it. The goal is to burn on an average rotation of five to seven years, but many lands could use annual or biennial burns, and even a couple of missed years can mean the difference between containing or losing a wildfire or being able to reintroduce fire. But everyone recognizes that without the legal protection offered by the law they would be out of business, and without the prescribed burning that it encourages the land would be turned over to lightning, arson, and accident. Like Florida's land acquisition program, the system wobbles between aspiration and desperation.

By the 21st century Florida had established itself at the apex of one of three dominant fire cultures in the country and had evolved into a regional and national leader on matters pertaining to prescribed fire. Through a variety of innovations—the Prescribed Fire Act, the Coalition of Prescribed Fire Councils—its influence seeped throughout the Southeast and then the nation. Even more astonishing has been the role of the Florida Forest Service.

As with most state forestry bureaus it has responsibility for fire protection. Unlike most it also holds significant lands of its own to manage. And like only a few others, it has committed to prescribed fire. (The only comparable states, in fact, were neighbors influenced by the Florida example.) While the Prescribed Fire Act was possible because a community of shared interests campaigned for it, it required a broker to coax it into law and an administrator to oversee its provisions. The Florida Forest Service did that job.

Nationally, the federal government has built the infrastructure for fire management, as it did a highway system; and it was the Clarke-McNary Act that helped state forestry take root in Florida. But the federal presence in Florida was modest, and split among agencies. Most forests remained in private hands as did the vast bulk of burning. It was the Florida Forest Service that brought fire protection to the counties, and then worked to temper rural woodsburning into prescribed fire. Among states probably only California forestry can claim comparable clout, but CalFire is almost exclusively a fire suppression operation, and it has little reach beyond its legal borders. The Florida Forest Service does it all.

And that, finally, is what accounts for its unique status within the national landscape. It suppresses fires on a scale few agencies match; probably none can equal its overall density of fire operations. It burns more acres than any other state, and has done so ever since record keeping began in the early 1980s. (There may be other states such as Oklahoma with as much private burning but they have no regulation over the process or hard data.) What made the FFS unique, however, is the way in which it serves as the connective tissue among Florida's various fire operatives. It was unusual for foresters to rally behind prescribed fire in the early years, and it is rare for a state to lead a consortium like that which has clustered around Tallahassee. The Florida Forest Service made both transitions. The kind of leadership role the U.S. Forest Service assumed nationally, the FFS has accepted for Florida, but without a hint of the former's old push toward hegemony. It remains first among equals. It works with the larger Florida fire community to establish standards, it issues the authorizations that put working fire on the ground, and it promotes that achievement through regional and national fire conferences, some of which it hosts. Neither the Feds nor private landowners can command such authority.

Within the Florida fire community, the FFS is widely recognized as providing entry-level hires for fire. Somewhat to its dismay, many of its cadets then go on to work for other agencies. So, too, the FFS has served historically as the entry point for modern fire management. Its policies and practices changed, often under duress, pushed by necessity as often as it was pulled by opportunity. But over the decades it established itself as the institutional fulcrum of Florida fire, without which the state's swarm of a fire community would struggle to leverage their ideas into practice.

FIGURE 1. Prescribed Fire Training Center. The real work occurs outside its walls, which was its purpose and the metric of its success. Photo by author.

FIGURE 2. Florida's interface—woods, exurbs, and flame. Photo courtesy Merritt Island National Wildlife Refuge.

FIGURE 3. Burning on the Choctahatchee National Forest, an earlier avatar of Eglin Air Force Base. Note the longleaf scored by turpentining and the raking around its base.

FIGURE 4. A landowner's right to burn. Most of the land burned in Florida is on private land, much of that still from ranchers. The Nature Conservancy also burns extensively on its lands and in collaboration with partners. Here the two meet. Photo by Parker Titus, courtesy the Nature Conservancy.

FIGURE 5. The wildland-galactic interface at Merritt Island. The streaking columns in the rear are convective plumes from prescribed burns. Photo courtesy Merritt Island National Wildlife Refuge.

FIGURE 6. Big Prairie at Myakka River State Park. The landscape as it more or less once was and as park managers try to retain or re-create, mostly with fire. Photo by author.

FIGURE 7. Big Cypress National Preserve. Pine, palmetto, rough, and grass—the classic surface fuels of Florida. Note scorch marks on the trees. Photo by author.

INTERLUDE

THE MANY REASONS FOR AND SINGULAR REALITY OF FLORIDA FIRE

When in doubt, tell the truth.

—MARK TWAIN

ON MARCH 12, 1935, a passing cold front drove a storm front of flame across 20 miles and 35,000 acres of the Ocala National Forest in some three to four hours. The saltating flames stopped only when they struck Lake George.

A typical spring wildfire had bulked up on steroids and it was obvious to everyone on the scene what kind of biotic steroids were involved. An overage scrub of sand pine and rough oak had woven a low, close-packed canopy of needles and leaves through which fire might spread as it would through a coniferous chaparral or a stunted old-growth patch of lodgepole pine. Put that fuel array anywhere and it would burn with unquenchable ferocity. In fact, much the same happened on the Osceola National Forest in 1956, when 100,000 acres of comparable combustibles blew up, and amid immature jack pine in 1980 at Mack Lake in Michigan that roared over 38,000 acres in an afternoon, or in genuine if overripe chaparral that powered fires across the Transverse Range over and over (Malibu also burned in 1935). All such biotas were prone to conflagrations: if they burned, they flashed in gulps through the canopy.

The Florida fire—variously named the Ocala, the Big Scrub, or the Bear Hammock—got ample attention. It put into flaming relief the argument that, amid such fuels, fire control was impossible. When fires in these settings burned, they burned so savagely that artificial firebreaks, even the Ocala-Daytona Beach Highway, were worthless. The only checks were an exhausted wind or a giant body of water. The best way to halt such fires was to keep them from starting or to knock them out while they were small. Prevention seemed hopeless in the face of an insurgency of woodsburners. Rapid attack was crippled by the poor infrastructure of a sparsely populated state. The only solution was to quench the fuels that stoked conflagrations. But dampening fuels before a spark struck begged the question of just how to prune them into a manageable state.

With big ideas, as with big fires, timing is everything, for several critical factors have to converge to make a perfect fire. The early 1930s did to forestry what they did to farming: monster fires spewed smoke palls that seemed to rival the dirt storms of the Dust Bowl. From 1932 to 1934 behemoth fires had blasted Oregon, the Northern Rockies, and Southern California, and raged across the Southeast. Stopping them had become a matter of national concern. On the public lands of the West, fire officers believed that rapid attack might work, and a month after the Big Scrub fire, the U.S. Forest Service announced a universal standard for fire protection across the country, the control of every fire by 10 a.m. the day following its discovery. Whatever the setting (or however remote) fire control could—would—attack the source rapidly at its origin. Every big fire began as a small one. An all-out, take-no-prisoners approach, backed by the muscle of the Civilian Conservation Corps (CCC), could end the nationally embarrassing big fires by throttling them in their backcountry cradles.

The Southeast thought differently. Its fire officers insisted that only regular, controlled fire could batten down the rabid fuels that powered outbreaks of wildfire. They could fight fires in a little-scrub landscape; they couldn't in a big scrub. Already researchers were experimenting with controlled burning for silviculture as well as for fuel reduction. In 1942 R. M. Conarro coined the term "prescribed fire" to indicate that fire could be targeted—could be scientifically directed—to distinguish useful burning from the generic woodsburning that, to foresters' eyes, plagued the landscape. As with medicine each prescription had a specific purpose. Hazardous fuel was the point of overlap between the reductionism of science and the simplified targets of management. In this way several interests converged: those who wanted to control conflagrations, those who thought that the best way to reduce fires was to reduce their fuels, and those who believed burning could be an acceptable technique if it was legitimated by quantitative science. Prescribed burning could count only if it was countable.[1]

Actual practice came more fitfully, often under the guise of "administrative experimentation." In 1943 an alloy of severe drought and unpruned rough reached critical mass and blew up on the Osceola National Forest to between 70,000 and 80,000 acres. This time there was no longer a CCC to muster any semblance of a fight against it. Meanwhile, two Forest Service researchers, C. A. Bickford and John Curry, had devised a guidebook with prescriptions by which one could reduce hazards. The burn and booklet were enough to entice Chief Forester Lyle Watts to visit the scene and then, although still mystified, to grant the Florida National Forest an exception to the national prohibition against controlled burning. Floridian foresters could burn for fuel reduction as the next and perhaps only solution to Florida's extraordinary circumstances. Fire control, fuel reduction, and field science all agreed that burning hazardous fuel was worth a phalanx of tractor plows and shovel-wielding CCC camps.

Revealingly, among the primary concerns to Bickford and Curry—because it was the "least tangible"—was the potential influence on public opinion. To advocate for controlled fire was "a direct contradiction" to the intense onslaught of fire prevention propaganda. "For foresters to now make use of fire appears, to many who have actively worked on these campaigns, a distinct breach of faith and a reversion of policy which may result in discouragement and confusion." With noble intentions state-sponsored forestry had promulgated a zero-tolerance policy toward fire, one many no doubt felt ill advised but necessary. The firefight was the drug war of its day. Now they had to face the consequences of not only admitting their error but of their willingness to push it anyway to promote a larger agenda.[2]

Even today, raise the issue of prescribed fire, and Floridian fire managers will recall the Ocala burn. It was, in a sense, the *experimentum crucis* for the proposition that fire would happen, and that the only strategy to contain it is to substitute tamed for feral flame. Fuel became the index of potential fire. As it sprouted upward, it would eventually reach the inflection point at which benign fires metastasized into malignant ones.

The arguments for hazard-reduction burning have never abated. Prescribed fire became a treatment of choice throughout southern pine plantations, the cleaned-up kissing cousin of folk woodsburning. When, during the 1960s, enthusiasm for restoring fire spread beyond the region, fire for fuel abatement remained a prominent justification, particularly among foresters. Such burns assisted the Sisyphean task of fire control. They helped shield foresters' trees, they dampened the slash left by the postwar's cancerous clear-cuts, they were sustained and guided by science. They looked like silviculture by other means. They remade what many members of the public, especially newcomers, might view as an eccentricity that bordered on madness into a form of security for life and property. Fuel reduction remains a primary purpose of prescribed fire nationally.

Yet it is only a part of Florida burning. Historically, promoting fauna, not reducing flora, was the principal motive for widespread burning. When observers commented on Florida burning twice a year, they were referring to pastoralists. More recently, wildlife has dominated the rationales. The justification for burning is to support habitat for endangered species like the Florida scrub jay, Florida panther, and red-cockaded woodpecker, for favored game like deer and bear, and for bird-watching. Most burning remains in private hands, even if done for a public purpose (think the Nature Conservancy). The purpose is not to get the fuels down so much as to get the land right. In most of the country, fire officers argue for fuel reduction burning as a means to get the ecology right. In Florida some fire managers are more likely to argue for ecological burning as a means to get the fuels right.

It's an important distinction. In most of the country the Endangered Species Act works to fetter prescribed fire; in Florida, it works to promote it. In the West pastoralism likes fire to remove brush, but won't tolerate routine burning because it will reduce stocking. In Florida pastoralists kept fire on the land for centuries, and continue to do so where possible. Elsewhere, private landowners largely shun burning as too dangerous and its smoke too wayward or likely to sink into settled valleys. In Florida they do the bulk of the burning. The National Fire Plan and healthy forest legislation made fuel reduction both a means and a metric.

Florida burning is too varied for such simplistic accounting: fire is an all-purpose ecological catalyst that stimulates fodder, recharges habitats, improves aesthetics, and along the way dampens fuels.

Besides, in a landscape so lush and volatile with combustibles, fuel reduction has limits. Even regularly burned landscapes will reburn from wildfires. Around Osceola wildfires rolled over land in 2007, 2011, and 2013 that had recently been burned under prescription. Fuel reduction burning can help soften their ferocity—make them easier to control—but a place that can burn twice a year won't allow one fire to stop a second. If fuel reduction is the means for burning, land managers may be told to find another tool that doesn't have the same risks or loose smoke into the sky. Internal combustion by wood chippers might replace open burning by driptorches.

Nationally, the emphasis on prescribed fire and mechanical treatments to help contain wildfire is suffering blowback. The fire problem has been redefined as a fuel problem. Funds available for ecological burning are being shunted into hazard reduction projects and removed from the backcountry to the wildland-urban interface. Fuel specialists replace fire ecologists. The narrative emerged that the problem with fire suppression (or more properly, fire exclusion) was that it led to a countryside obese with combustibles. The argument for prescribed fire has narrowed into a plea for fuel reduction. What should have been a three- or four-legged stool has narrowed to one unstable leg.

This is blowback. The federal fire community finds itself exactly in the quandary that worried Bickford and Curry in 1942, that reversing arguments would be seen by cooperators as a "breach of faith" that might result in "discouragement and confusion." All of that has again come to pass. Of course there remain plenty of places, especially around the built environment, for which fuels management is mandatory, and there are others in which fuels serve as an interim index for fire's restoration; if managers get the fuels right, fire's ecology can sort out the rest. But the worry is that the fuels argument, like fuels, may create a backlog and buildup problem. The toxic residue of bad-faith arguments, like the legacy accumulation of fuels from the poor practices of the past, may well make contemporary work more difficult.

This is not a problem for Florida. Fire is so essential, so inevitable, so pervasive that you just need to burn. On one site the burning might be primarily for fuels; in another, for cattle; elsewhere, for scrub jays, saw grass, or gopher tortoises. No single reason prevails because in Florida fire is as fundamental as rain. It just needs to happen. The choice is whether it will happen through nature or with some guidance by people. The choice is between seasons, frequency, and maybe intensity, not whether or not the land will burn. In Florida the reasons for burning are many, but the singular reality is that fire will come.

PANHANDLE AND PENINSULA

THE MORE THINGS CHANGE

Eglin Air Force Base

THE CHOCTAWHATCHEE WAS ALWAYS a marginal landscape, its woods stocked with longleaf but not densely, its sandy soils too depauperate for cash crops, and above all, isolated. Its inhabitants lived in a subsistence economy of hunting, fishing, kitchen gardens, and open-range herding of cattle, sheep, and hogs. Rivers took its commerce, stream by stream, to the Gulf, leaving each corridor isolated from its neighbors. Even the Pensacola and Atlantic Railroad, rewarded with alternate-sectioned lands, could not break the pattern. There was a market for the Choctawhatchee yellow pine, both as saw wood and as naval stores, but bulk transport of timber was costly, which argued for distilling the longleaf into turpentine. It was the same logic that drove pioneers to ship whiskey rather than cobbed corn.[1]

So when President Teddy Roosevelt set aside the Choctawhatchee National Forest by proclamation in 1908, there was plenty of public land still unpatented, although private holdings, not all occupied, dappled the gazetted 133,000 acres. The primary economy hinged on the longleaf. When the big trees were stripped away, the focus shifted to turpentining, or what the Old World had known as resin tapping. It was the technology of rubber tapping applied to pine-laden pitch in which the trees were scarred, the sap collected into "boxes" or cups, and then boiled off into distillates. It was a crude, brutal practice, as damaging to transient workers or African Americans trapped into debt peonage as it was to longleaf. When the sap failed, the trees were felled and hauled off to mills. What remained was a landscape of low-grade pine and turkey oak. The biota showed the same tough fragility as the sand hills.

With quickening pace the practice self-destructed on the private lands, effectively scalping "orchard" after orchard in a kind of slash-and-burn operation, while the Forest Service gradually tamed turpentining on the lease lands under its jurisdiction, laboring especially to encourage regeneration. Meanwhile cattle grazing faded out during the era of tick eradication, and sheep and swine departed in the 1930s. Over the decades the Forest Service bought those degraded and abandoned lands as it could, both those within the forest boundary and those adjacent. During the Depression many holdings fell into tax delinquency. By 1940 the Choctawhatchee forest had nearly tripled in size to 340,890 acres. But even as the forest grew, the domain of the longleaf shrank. As a working forest the Choctawhatchee was played out. The Forest Service sought rehabilitation by planting with slash pine and instigating fire control.

It was with mixed feelings, then, that the Forest Service allowed the forest to be transferred to the War Department in 1940 as a proving ground. The military wanted a sparsely populated coastal area that it could bombard from air and sea. In 1935 the Valparaiso Realty Company had donated 1,461 acres around Choctawhatchee Bay for that purpose, what became the Valparaiso Bombing and Gunnery Base, and in 1937 it was renamed Eglin Field. The 1940 transfer expanded that beachhead on a scale to suit the coming war. Before the handover, most of the remaining salvageable pine was cleaned out.

The Choctawhatchee would thus be reborn, after a baptism of fire, into a new destiny that seemed no less destructive than what it was leaving. A trashed landscape, it seemed, would be slashed and burned on a war footing. Yet that perception misread the landscape's legacy. It retained the character of a national forest as much as it acquired the identity of a military base, and it remained a southern landscape as much as a national forest. In odd, even eerie ways, it swapped the axe for ordnance, but it held to its older traits. It continued to burn.

What the precontact fire regime might have been is unclear, only that there was ample fire. The Panhandle remains today a hotspot of biodiversity; an estimated 25 percent of all the plants in North America crowd into the longleaf ensemble, and within the wide-ranging longleaf, the highest concentration lies between Tallahassee, St. Marks National Wildlife Refuge, and Pensacola. The cartography suggests the area as a long-surviving refugium for longleaf, or to use an old anthropological expression, one especially appropriate granted the abundance of fire, a hearth.

The density of humans in a place bears no more relationship to the density of fires than the simple density of lightning strikes does. The right kinds of sparks have to strike the right kind of kindling. Nor does the number of starts translate directly into area burned. People move, fire propagates—small bands who circulate around a landscape and burn patches in seasons account for far more burned area than a higher population that cultivates those patches intensively. Once loosed on a landscape a few fires can linger, creep, and spread over large areas. So there are no simple algorithms based on lightning or populations by which to recreate the regime before the hard record of fire scars and written documents. There is only the ecological legacy that says there was a lot of burning.

Shortly after the Choctawhatchee became a national forest, its supervisor, Inman Eldredge, wrote a detailed description of its fires and how they defied the fire-control directives of the national program. It's one of half a dozen classic summaries of traditional burning available globally. Eldredge depicted the people "right down on the ground, the settlers, the people who lived in the woods, the turpentine operators, and so forth," as "the greatest, ablest, and most energetic set of woods-burners that any forester had to contend with." His account aligns well with such other sources as the early reports from Herbert Stoddard and Roland Harper. It sketches the last era before the great transformation felled the longleaf across the region and scattered the flames it needed for renewal. It's worth quoting at length.[2]

The Choctawhatchee Forest, Eldredge argued, was unique. The methods used in the western forests "seldom" had value when applied to it, and then only after severe modifications, and nowhere was this more "clearly shown" than "in the matter of fire protection." Some 60 percent of the forest remained in private hands, largely from railway grants, so it was impossible to keep people out ("nearly all native Floridians of the 'Cracker' type") or to contain their fires.

The popular sentiment of the residents within the Forests, in common with nearly all of the people of the South, is unqualifiedly in favor of the annual burning over of the pineries. The homesteader and the cattleman burn the woods to keep down the blackjack, undergrowth and to better the cattle range. The turpentine operator burns over his woods annually, after raking around his boxed trees, and at the time when the burning will do least harm in order to protect his timber from the later burnings that are sure to occur. He burns also with the idea of keeping the turpentine orchards clear of undergrowth and free from snakes, in order that the Negro laborers may gather the gum with ease and safety. The camp hunters, of whom there is a large number during the fall and winter months, set out fires in order to drive out game from the thickets. All of these different classes of people have for a great number of years been accustomed to burning the woods freely and without hindrance of any kind, and it is done without the knowledge or the feeling that they are breaking the laws or in any way doing damage. On the contrary, they all have the most positive belief that burning is necessary and best in the long run.

The turpentine operator burns his woods and all other neighboring woods during the winter months, generally in December, January, or February. The cattleman sets fire during March, April, and May to such areas as the turpentine operator has left unburned. During the summer there are almost daily severe thunder-storms, and many forest fires are started by lightning. In the dry fall months hunters set fire to such "rough" places as may harbor game. It is only by chance that any area of unenclosed land escapes burning at least once in two years.[3]

This "annual burning" had gone on so long that the fires did little harm: it was their absence that upset the local ecology. Attempts at fire exclusion were "disquieting" because "each year of protection from fire makes the fire danger greater." Let a patch stay unburned for six to eight years and a fire will result "in almost complete destruction." On the contrary, the practical ecology of burning made perfect sense. The fires were a kind of sweeping, vacuuming, and rug beating that kept the ecological household habitable.

Longleaf and Cuban pine apparently suffer but little, if any, from the light annual burning of the woods. Even seedlings as young as three years old

> appear to suffer no damage, although there is no doubt that the youngest seedlings seldom escape destruction, and the poor reproduction of pine is largely due to the frequent fires that keep the seedlings from getting a start. The undergrowth of oak is burned to the ground, but coppices so profusely that an annual burning is considered necessary to keep it down. The forage grasses are, of course, burned to the ground but recuperate very rapidly and seem to suffer but very little. In fact, the stockmen claim and believe that the grass is much benefited by the light annual burning.[4]

Without fire this way of life was impossible. Annual burning was, Eldredge concluded, "the universal belief of the Forest residents" and most of the landscape "burned over each year as a matter of business." There was no perception of damage because the yearly burns ensured there could be no damaging fires. Fire and forest existed in rough accommodation. But if something broke that equilibrium, if turpentining ended, if widespread felling stacked up woody combustibles, if ticks compromised cattle, if sheep and hogs left, if the Depression brought even the subsistence economy to its knees and land fell tax delinquent, if foresters acquired the lands and replanted them to slash pine, then the old ways no longer worked as before and might themselves add to the mess.[5]

And that is what happened. Most of the old pieces still remained, although in different proportions and with different responses to fire. Oak became entrenched; longleaf struggled to revive; rough spread like continually broken scabs; fires beat to atonal rhythms, even damaging what they had once preserved or healed. The Forest Service struggled mightily, assisted by the CCC, to beat down fires and raise up a new crop of planted pine, with mixed results. Critics thought that fire control was actually retarding the effort, that slashing without burning only yielded the terrain to oak and scrub, that hogs and mast would replace native species and longleaf. But foresters, trained to the long view, were ever hopeful, and many believed that rehabilitation was at hand. Then the Choctawhatchee went to war.[6]

As World War II slipped into a long Cold War and assorted hot ones, the military mission at Eglin grew, acquiring new roles and "tenants" much as the national forest had handled leasees and private inholders. The Air Force made the facility its primary site to plan and conduct various testing programs, from missiles to Agent Orange to smart bombs and drones. In 1990 it was redesignated the Air Force Development Test Center to test, evaluate, and support the "development of conventional non-nuclear munitions, electronic combat systems, and navigation/guidance systems." By the end of the Cold War some 60 groups operated out of Eglin, from Army Rangers to Navy downed-airman trainers. It was a multiple-use air base.[7]

The Air Force remade the Choctawhatchee as the Forest Service had the lands gazetted to it, and as before much of the earlier imprinting endured. The reserve continued to harvest timber (it had its own sawmill until 1974); boosted conifer planting, shifting from slash to longleaf; allowed hunting, hiking, and recreational camping; and in the early 1970s blocked out 220,000 acres for two "environmental zones" for the encouragement of indigenous species, a Department of Defense (DOD) alternative to wilderness. To support its military mission some 66,000 acres were cleared for test ranges, roads, base facilities, and other needed infrastructure. Habitat rehab relied on mechanical slashing, herbicides, and fire. Even with clearings, Eglin oversaw the largest forest under DOD administration. The outcome looked a lot like a southern national forest, save for two firing ranges, but even these sacrificial lands might be likened to the phosphate mines that scarred the Ocala.

One reason is that the Sikes Act of 1960—the same year as the Multiple Use-Sustained Yield Act for the Forest Service—committed the Defense Department to "provide for the conservation and rehabilitation of natural resources on military installations." As with the national forests, DOD lands became subject to a succession of environmental legislation that put it under similar regimens as other public land held "in trust." In a moment of euphoria Eglin even advertised itself as a "proving ground for conservation." The critical moment for Eglin came with the Endangered Species Act, for the reserve held the largest extant patch of old longleaf pine in the country and hence one of the prime habitats for the now legally protected red-cockaded woodpecker, which was the F-16 of southern environmentalism. Eglin's indigenous firebird was flying into a collision course with its warbirds.[8]

The controversy was defused. It would be bad publicity for DOD to pull rank on the Endangered Species Act, and it would be unwise for the Fish and Wildlife Service to appear to hamstring national defense. Moreover, DOD had ample monies to sponsor detailed studies about how to sustain and even improve that critical habitat. In 1994 it commissioned a seven-year inquiry into natural resource management from a consortium of researchers drawn from the University of Florida, the Nature Conservancy, and Tall Timbers Research Station. Nineteen days after the 9/11 terrorist attack, the group presented its final report, centering on a holistic approach to longleaf restoration. Still, even at Eglin, not everything could be done everywhere. Management had to rank goals; and they chose to focus principally on sandhills longleaf, the prime habitat for RCW. Unsurprisingly, fire was close to the core.[9]

It had always been a fire forest, and fire management would determine the future of its land management schemes, just as Inman Eldredge had early recognized. Users changed, incendiaries changed, the purpose of burning changed, but fire and land remained inextricably intertwined. Much of the ordnance tested could start fires, and many were designed outright as incendiaries; among the early experiments were tests that had assisted the firebombing of Japanese cities. Fire control improved accordingly, particularly during extensive reforestation efforts in the 1950s. But fire's removal—not just by actively suppressing flames but by no longer lighting burns virtually year-round, year after year—had already damaged the biota. Eglin's reformers might want to push ahead, but they first had to undo the past, which could never be wholly remade, and they spent as much time defending their rearguard against the ill-conceived if well-intentioned acts bequeathed them as advancing into their desired future.

The recovery of fire matched the recovery of the longleaf. In some respects military oversight helped. Commanders were accustomed to identifying priorities, isolating tasks, committing resources to meet goals—and DOD had resources that few civilian agencies could match. The effective fire control that the Forest Service had begun with CCC labor the Air Force could continue and amplify. Wildfire had burned a scant 267 acres a year from 1933 to 1937. Then, as forests regrew or were planted, burned acreage began to climb, and the character of the burning morphed with the change in woods, which made routine surface burning trickier. In 1931 the forest had 7,000 acres of sand pine; by 1979 it had 60,000. Even as it strengthened firefighting capabilities, Eglin escalated a program of winter controlled burning.

But the era of easy annual burning by local residents was over. As lightning, arson, and ordnance set fires, they smoldered or blew up, and environmental constraints from threatened and endangered species, liability and permitting, and smoke all restricted free-ranging fire. Wildfires threatened to swell and prescribed fires to shrink. Moreover, prescribed burning struggled against a backlog of decades: there just weren't enough burn days and the flames didn't drive the scrub back. The prescribed burns were, in a sense, a strategic retreat, holding and occasionally yielding, but not counterattacking.

Still, as it commissioned its research, the Eglin fire program also experimented with new field tactics. A handheld driptorch could ignite 250 acres a day. But torch-equipped ATVs could increase that acreage by a factor of three or four, and ignition by helicopter could boost it an order of magnitude. Meanwhile, ordnance-kindled fires demonstrated that summer fires could stimulate Eglin's mix of bunchgrasses into seeding, and hence spreading, and could burn hot enough to hold against hardwoods. Where restoration was the objective, mechanical cutting and Velpar herbicide were used as a force multiplier. By such means Eglin began to recover the cycle of year-round burning.

All these practices were tactical innovations. They played out against a strategic change, what might be termed the environmental equivalent of nation building. Fire officers understood that landscape restoration was a decades-long commitment; that the landscape demanded fire; that fire depended on fine fuels—wire grass and needle cast—that would need many years to build up and spread. In brief, they accepted that they could not do everything at once. They had to prioritize Eglin in a kind of battlefield triage. There were places they could easily hold through regular burning. There were other sites that, with steady labor, they could remake into usable habitat. And there were places that were too close to towns or facilities or would smother the region in smoke or were too marginal to the mission and would be left to wildfire and the future.

Not least, Eglin joined a larger coalition engaged in restoring longleaf, RCWs, and fire. The Gulf Coastal Plains Ecosystem Partnership pooled and orchestrated the fire efforts of 10 agencies, a growing quilt of nearby national forests, state forests, water management areas, nature preserves, and Nature Conservancy holdings. That bonded the land side. The fire side alliance strengthened as DOD declared as policy that its bases would adhere to the 1995 federal wildland fire management policy, would sign the Interagency Fire Management Agreement, and would station a liaison at the National Interagency Fire Center. Eglin would muster its resources with those of its neighbors, and it would join the national consortium of wildland fire agencies. If its lands looked a lot like a national forest, its fire programs looked a lot like those of the Forest Service and Fish and Wildlife Service.

All in all, it seemed an odd historical inversion. Foresters had long likened their fight against southern woodsburners to an insurgency, which they steadily subdued with a battery of suppression, prevention programs, and land use changes. Now they realized that the fight had not been over fire itself, but over who held the torch and to what purpose. They sought to substitute good fires for bad ones, as they might seek to replace poppies or coca with more benign crops such as wheat or apples. They set prescribed fires to forestall wildfires, they conducted programs of public education, and they sought to alter the character of the land to support their larger mission. A century after the Choctawhatchee was reserved as a national forest, the fire program was burning an average of 100,000 acres a year.

It seems incongruous that a place committed to the destructive use of fire—weapons testing—should pioneer in fire's constructive use for habitat restoration, and that a military base should commit to the national security of a woodpecker. Yet the oddity is easy enough to explain if you recognize the fundamental character of the site and the imprinting of its human heritage.

The persistence is striking. Today, 60 percent of Eglin's 463,448 acres is open to public use; this is the same percentage of land in the original Choctawhatchee forest reserve that was private. Similarly, fire practices have modernized but persisted. Certified burn bosses have replaced woodsburners; bombing has replaced logging; and the protection of RCW nesting trees has supplanted the "boxing" of longleaf for turpentine. In both cases, standard practice calls for raking around the trees and then burning off the surrounding wire grass and rough. The paradox, however, easily dissolves into another illustration of the Florida fusion.

At Eglin the Florida formula is in full display. The land is an ecological palimpsest, ever recycled to new purposes but with the old inscriptions only overwritten, never wholly erased. Likewise the fire regime continues to reassert itself. The old flames had been rubbed almost away but then reappeared through the parchment, though in a more modern language; instead of medieval Latin overlaying ancient Greek, ecological burning overlay turpentiners' protective burning. The recovered fire regime went back to the future, through many months for many purposes. Perhaps most strikingly fire here has been a point of union rather than secession.

What elsewhere split groups, here joined them. In the West an endangered species typically forced a halt to burning. At Eglin, as throughout Florida, it supported it. Eglin's warbirds set fires, its firebirds needed a burned landscape, and fire managers had to see that they aligned. The Department of Defense, the Fish and Wildlife Service, the Florida Forest Service, the Nature Conservancy, environmental nongovernmental organizations—all might argue over the best use of the land, how to restore a critical habitat, and what regimen of burning was ideal, but they all agreed that fire was essential. Whatever happened, there was no dispute that prescribed burning would be the common catalyst.

The heart of the Eglin test site is a long flight path of 10,000 acres that is routinely blasted and burned. Crews annually fire the forests around that swath to prevent escapes. Those woods may be the most heavily burned longleaf in the nation, and paradoxically they may most closely approximate the fire landscape that flourished prior to its wreckage through logging and turpentining. Most of the testing concentrates on the southern end of the range. And it is here that the red-cockaded woodpecker has its highest density of nests.

REGIME CHANGE

St. Marks National Wildlife Refuge

THE PROCESS OF ASSEMBLING the biota of what became Florida was less a mosaic, which implies a preformed pattern into which the pieces fit, than a kaleidoscope that jumbled them in ways that could be viewed by observers as orderly.[1]

People and climate each had a hand on the device. Sometimes they twisted the same way, sometimes sideways, but always with fire in the viewer. Around the St. Marks River Spanish contacts in the early 16th century found an Apalachee culture that had blossomed into extensive agriculture and mound building. Soon after Pánfilo de Narváez and Hernando de Soto left, the indigenous population began a horrific crash, probably under the blows of introduced diseases, that swept perhaps 90 percent of the human presence off the landscape and left fields to fallow. The kaleidoscope turned; and it kept turning through wars and land use schemes that chewed, plowed, abandoned, turpentined, and occasionally slashed through the scene, though always with fire in the mix. Along the coast residents also fished, hunted waterfowl, and smuggled. Permanent structures were few; building materials were sparse, transportation spotty, and hurricanes common. What endured was the complex bay, its marshes, and the adjoining woods. By the 20th century the Tallahassee region south toward the sloughs of the St. Marks River had become, paradoxically, a hotspot of biodiversity.

In 1931 53.2 acres were set aside by presidential order to promote waterfowl. Two years later the St. Marks Migratory Bird Refuge acquired a CCC camp and began adding land along the shore and inland that would, over several decades, swell to 67,563 acres and absorb most of Apalachee Bay and its hinterlands. The new holdings included barrow pits, pine plantations, cornfields, dikes, and a "tram road" used by loggers. Lands not in the refuge proper often slid into the Apalachicola National Forest, established in 1936, or other public holdings, such as the Wakulla Springs State Forest and the Ochlockonee River State Park. The interstitial landscape was largely exploited by timber companies, of which the St. Joe Paper Company held 1.5 million acres. In their way they were all working landscapes, built on a common wreckage of recovered (mostly planted) slash pine, and had a shared experience with fire. What happened on one could easily cross shared borders. In 1975 Congress designated 17,746 acres on the refuge as the St. Marks Wilderness and 24,612 acres on the Apalachicola as the Bradwell Bay Wilderness. In the greater St. Marks area more private land migrated into public land, and the various public agencies moved closer in policy and practice.

In 1997 the St. Joe Paper Company became the St. Joe Company and joined the "irrational exuberance" that had spilled out of the stock market into real estate. Its lands were more valuable for growing houses than slash pine. A bevy of state, federal, and private conservation groups intervened to buy the land before sprawl could. The St. Marks proper was authorized to add 35,000 acres. But much of the buffer lands was slowly purchased or braced by conservation easements. In 1997 some 60 percent of the refuge was wetland and 40 percent upland. The expansion would equalize that ratio, and if the neighboring estates were figured into the sum, the story moved from a marshy beachhead to an expansive flatland. Usefully, the expansion coincided with the advent of the National Fire Plan. In practice fire crews had become interchangeable, guided under a common protocol of burning and purposes. By 2010 St. Marks National Wildlife Refuge (NWR) was authorized at 109,764 acres, which represented nearly every habitat in north Florida. Almost all of those lands were burnable. The issue was how to burn them.

As the landed estate had evolved, so had fire practices. From its origins the marshes burned. Ranchers, hunters, and trappers had seen to that, and the early staff left the practices alone or co-opted them as they gradually assumed control over timber, grazing, hunting, and other traditional pursuits that stirred the customary with the quirky (such as worm grunting, in which patch burning was used to assist harvesting worms). Distinctively, St. Marks was the first federal refuge to receive official acquiescence for its burning.

The old ways had been outfitted with a uniform, but otherwise continued as before. Old timers recalled striking matches against the door of a pickup and tossing them to the roadside. They would determine fuel moisture by bending pine straw. Often they would burn late in the afternoon so that patchy fires would self-extinguish as evening humidity rose and congealed into dew. In 1941 John Lynch of the Fish and Wildlife Service published a review of marsh burning along the Gulf Coast. He identified three types of burns, each useful to different ends, and each potentially damaging if done poorly. The *cover burn* was a flash fire through grasses over standing water; the *root burn*, during dry spells, consumed roots that had grown into litter left unburned for a number of years; and the *peat burn*, as rare as deep droughts, scoured out the organic soil altogether. Each in its way was essential. Without cover burns cattle and geese would go elsewhere; without the occasional peat burn the marsh would fill in and vanish. "In fact a large part of the best waterfowl habitat in this region is the direct result of deep burns."[2]

Fire would happen: how it happened affected different species. Geese favored clean burns; marsh ducks, shorebirds, and fur-bearing mammals sought out spotty burns. Cover burns in the spring promoted forage and trapping, but also afforded some protection against summer wildfires from either lightning or gator poachers. The pattern of refuge burning was accordingly diverse. Staff burned in winter and spring, day and night, wetland and upland, hot and cool as useful; firing techniques varied with species, nuances of sites, and local skills. The upland forests, still replanted and recovering from their gouging and scalping, had to integrate grazing with slash pine reproduction. The woodlands became another suite of species with their own concerns over burning severity and timing; they too had their nesting areas in need of protection and their competition against unburned roughage.

By the early 1940s what was becoming known as "prescribed burning" laid down guidelines for forestry. Fire management assumed the shape of silviculture, as pine plantations became the latest of southern row crops, after corn and cotton. The more plowed lines the better the control, so burners dragged driptorches between them like sowers or hoers, ideally in the winter and always with a steady north wind. C. A. Bickford and John Curry codified the technique into a system with their 1943 paper. A dozen years later the Florida Forest Service publicized the experience with a pamphlet *Using Fire Wisely*.

Like a crop replanted year after year, however, the pyric system steadily degraded under the continual repetition. But it was a technique that could coax the inherited slash pineries into production and it was something the refuge's "militia" fire crew could do. Frank Zontec, a forester arriving from Kentucky Woodlands National Wildlife Refuge in 1965, recalled the ragtag band of maintenance workers, equipment operators, biological technicians, and managers who turned out to do the burning with a "military surplus tractor plow unit, a couple of homemade drip torches, a couple of military surplus back pack pumps, shovels, and fire flaps." That, as he recollected, "was it."[3]

The fire revolution of the 1960s affected St. Marks differently than most places because it already had an officially sanctioned burning program. The first Tall Timbers fire conference included a contribution from the Fish and Wildlife Service that spotlighted St. Marks. Of more significance than the Leopold Report or Wilderness Act, both critical in restoring fire to the West, was the Endangered Species Act (1973), which boosted the political power of the Fish and Wildlife Service and, like an atlatl to a spear, gave leverage to agency burning. Still, know-how was local; not until the fatalities at Okefenokee and Merritt Island did the FWS create a countrywide fire program and join the national fire infrastructure. What did come out of the early decades of the fire revolution was a more robust science of fire's ecology and a shift regionally to more varied burning. Timber production continued to ease out of the picture and species habitat encroached, like osprey moving back into vacated land. Oddly, the refuge burned more in the woods and less in the marshes.

National and local reforms quickened during the 1980s. The refuge experimented with growing-season burns, with helitorches and then Premos that dropped combustible ping-pong balls, with looser responses to wildfires. As the Forest Service changed its mission, St. Marks and the Apalachicola began to merge into a common culture of fire. The row-crop model dissolved into mixed firing tactics in support of multiple fire ecologies. In 1989 the refuge abandoned its old silviculture altogether in favor of all-aged stands, and its fire philosophy jostled into an accommodation.

By the time Frank Zontek retired, St. Marks had a dedicated fire crew—outfitted with Nomex, trained to national standards and dispatched for national assignments, equipped with engines, ATVs, two tractor-plows, a contract helicopter, a spotter plane, and elaborate computer support. At a technical level the program could be interchanged with any in the state. From rural burners with uniforms, the staff had evolved into strict prescriptionists and then into full-spectrum fire managers who had largely dispensed with internal firelines, kept plows to the boundaries, allowed wildfires and themselves some room to maneuver, and encouraged mixed burns to mixed purposes.

By now St. Marks had become famous as a breeding ground for prescribed burners, part of a migratory circuit of fire folk as they flew east and west and from agency to agency.

Throughout, only the basic pieces had remained constant. All the arrangements and rearrangements of those pieces, even the pattern of their rearranging, had changed. More unsettling, so had the ideas by which those changes might be understood.

Fire, landscape, agencies—all had persisted and all had reformed often independently from one another. Landscape had changed because of economics and politics; the Fish and Wildlife Service because of the Endangered Species Act (ESA) and a catastrophe at Merritt Island; fire because, if wild, it absorbed all the changes around it and, if prescribed, it reflected people burning to new ends with novel means. The land had morphed ceaselessly, from abandoned turpentine orchard, cornfield, and woods pasture into slash pine plantation and then into habitat for nongame birds. The agency had matured from good ol' boys tossing matches into a sophisticated, high-tech program with geographic information system maps, helicopter-borne ignitions, and interagency standards for firefighter fitness. The politics had redefined goals from harvesting timber and nabbing alligator poachers to protecting critical habitats for flatwoods salamanders and whooping cranes. Its history had not followed a simple narrative arc but had turned kaleidoscopically. Among the pieces in the tumbler were ideas.

Ideas mattered because they did not simply record what happened but helped shape events. What lodged in the head could move the heart and direct the hand. People managed fire according to how they understood its role. When they believed fire to be intrinsically bad, they sought to suppress it. When they thought fire good, or useful, they strove to put it on the land. For fire fundamentalists fire was simply bad in and of itself. For fire pluralists, however, fire was many in its origins and effects, and they appreciated that plants and animals were not adapted (or not) to "fire," but to the patterning of fire, much as they were not adapted to "water" but to the rhythms of rainfall and flooding. The solution was the concept of a fire regime.

"Regime" (and "regimen") derive etymologically from the Latin *regere*, to rule, manage, direct. It entered Middle English through the Norman conquest and long maintained a political connotation (think of cognates such as *regal* or *regulate*). It was first used to describe fire in the context of 19th-century French colonial rule in North Africa and Madagascar (*régime des feux*) and occasionally within British colonial circles, whose foresters trained in Nancy, again with a political cast, as in the indigenes' "dictatorship over the ecosystem." From time to time it was rediscovered or reinvented. It was not applied to natural systems until the American fire revolution during the 1960s. Its modern usage dates to two papers by Australian Malcolm Gill in 1973 and 1975 in which he ordered (*regere*, again) such critical parameters as frequency, timing, intensity, and extent, and urged researchers to use "fire regime" instead of ambiguous expressions like "adapted to fire." Within a decade the concept had become widespread in the literature and was codified in the 1984 textbook *Introduction to Wildland Fire*.[4]

The fire regime concept boosted the precision and analytical power of fire ecology. But even as it clarified ambiguities, it introduced two difficulties of its own. One, it replaced the messiness of fire with a Gaulic logic. Each site has its own fire regime, and that regime was constant because it was inherent in natural conditions. For those intent on disciplining unruly woodsburning or those keen to restore fire, it explained the pattern that should prevail—the governing rule, as it were. Here was the second effect, that the concept harked back to its old rootstock in politics. It segued from describing into prescribing. By recounting, in principle, what pattern of fire had existed on the land (or had existed before European contact) it set norms for the kind of regime that ought to exist. It laid down a formula by which to burn.

In short order, fire regimes were themselves organized and ranked, much as silviculturalists did type classes. Class B referred to infrequent, low-intensity surface fire with more than a 25-year return interval—Eastern deciduous forest and pinyon-juniper woods in the West, for example. Class E characterized shorter to medium-length return interval crown fires or stand-killing, high-intensity surface and ground fires with 25- to 100-year return intervals, say, boreal forest or coastal chaparral. The purpose of fire ecology was to identify a site's fire regime, and the purpose of management was to maintain (or reinstate) that pattern. Such notions got encoded into fire management plans. A concept that had emerged to challenge simplistic and dangerous characterizations of fire had become itself not only simplistic but operational. Fire regime classes grew into the intellectual equivalent of row-crop burning. It's as though Clementsian ecology had returned from the dead.[5]

The concept is now seriously out of sync with what is happening on the land. It can't embrace the variety of fires that a place like St. Marks needs: refuge management is constantly outside the lines of what the concept stipulates. Either they multiply the number of regimes on the refuge, like taxonomic splitters making every local aberration into a new species, or they scrap the concept, like lumpers stirring everything into a common stew. If they let new regimes breed like rabbits, the concept loses meaning. If they dump the idea, they lose the intellectual order it brings. A third option of course is to reconstruct the concept into something closer to the landscapes of St. Marks.

That requires, first off, scotching the idea that a fire regime is something embedded in the geographic matrix of a place, that it is inherent and immutable, that the regime that existed at some arbitrary time is normative and ought to serve as a standard for management. The reality is that burning has constantly changed along with the land and with people's residence on it. Second, it requires accepting the fire regime as a statistical composite. It more resembles a climate than an orbiting planet, ceaselessly repeating its cycles. Fires occur within a regime much as storms do in a climate type; any regime can have a mix of storms just as a climate can experience a variety of storms. Yet there are clearly differences among places (or among times) in how fire occurs just as there are among climates. While they both display similar kinds of fires, the chaparral of the Transverse Mountains have a different regime than the Dakota prairies, much as Cleveland has a different climate than Phoenix despite mixes of drizzles and thunderstorms.

This is a condition of nature. Most of Florida would seem to have a simple and similar regimen of fire: it burns annually, or biennially, or at short intervals. But there are plenty of instances where fires went missing and the rough thickened and rather different burns resulted. Besides, there is the conundrum of exceptional events such as the floods that began 1998, the droughts that succeeded them and fueled conflagrations, and the four hurricanes that struck Florida in 2003 and scrambled its woods. The idea that fire might return (or cycle) in the ideal way foresters sought a rotational harvest of trees is nonsense. Still, rude patterns exist: despite some similar storms, no one would confuse Florida's climate with New Hampshire's; nor would they mistake Florida's fire regimes for those of Nevada.

It gets worse, however, because people have constantly interacted with the natural setting of the greater St. Marks region to alter fire's appearance. The Apalachee culture piling up mounds and cultivating corn. Turpentiners and loggers cleaning out the longleaf. Open-range ranchers loosing cracker cows through the countryside. CCC enrollees scraping roads and erecting dikes. Refuge managers contriving habitats for waterfowl. Each remade the landscape and set and stopped fires differently. And to this roster one should add researchers, as each generation queries the place with its own distinctive apparatus for analyzing and advocating.

The fact is, research is not neutral because it affects what it studies—this, after all, is the point. Nor did research questions emerge in situ. Just as markets far removed from the Florida panhandle created demand for naval stores or longleaf timber or egret feathers and affected land use on the refuge, so ideas that evolved in settings like the Nebraska plains or the Northern Rockies could influence what happened around St. Marks. They were transmitted not by sail or steam but by scientific journal or professional society. Such ideas influenced research, which affected policy, which affected land, which was then studied by other researchers testing their concepts against nature, which then influenced policy—an intellectual fire cycle, or management fire as a Möbius strip.

What fire regime characterizes St. Marks? Pick a time. Pick a place. Take a number. How should such a regime be characterized? With a lot more looseness than the concept now allows. Disturbance ecology abandoned decades ago notions that nature was in equilibrium, or if perturbed could be restored to that putative balance. Yet the fire regime concept has come to embody just such assumptions of stability. The landscape may no longer be viewed as a static scene, but the processes that organize it are presumed constant, so process preservation can substitute for scenic stabilization, and the fire regime concept orders one such process, fire. Moreover, the concept comes heavily laden with normative freight. If you know the proper fire regime, you must know how far the present regime has deviated from it, and what you must do to set matters right.

The fire regime concept was an enormous improvement over arguing about "fire" as a singular phenomenon. With time, however, it has itself become a singularity. It no longer captures the exuberance of fire as it appears on the land or the hodgepodge of practices needed for fire to do the ecological work required of it. As prescribed fire has evolved from row-crop burning into more free-form expressions, so fire concepts need to find their versions of the loose-herded wildfire, the spot-kindled burn allowed to poke and probe for weeks, and the prescribed burn held with wetlands and evening dew instead of plow lines. The ongoing fire revolution must reform its ideas as well as its policies.

It's easy to grasp why the fire community has seized upon the fire regime concept in its old form. It's easy to understand, it's simple to teach students, it suggests testable hypotheses, it gives clear guidelines for operations and standards against which to measure success. It's "legible" to authorities. Its only flaw is that, like Clementsian succession, it is becoming less useful as an explanatory system and more dangerous as a normative one.

What is the right fire regime for St. Marks? The question is not simply a matter of science but of politics and of ethics. It is, in fact, fire management's version of Aldo Leopold's land ethic, and it suffers the same liabilities as a program for operations because an answer is not embedded in nature but emerges from the interactions of people and land, and, we need to add, of how people conceive of those interactions. Of course refuge managers want to do the right thing. It's just not obvious what that is or even how to determine it.

The near future suggests that, regardless of what the science says or what fire regime is prescribed, fire crews will continue to experiment and do what seems necessary to promote the habitats the refuge requires. They will pull back from remote pine islands and let lightning fire work its magic. They will try some night burning. They will set summer fires to drive back titi thickets. They will prescribe burn in wilderness. They will mix alloys of flame, herbicides, and saws. Their "toolkit" will contain as many kinds of fires as their machine shop does wrenches.

It will be helpful if fire science can keep up. It might begin by revisiting the fire regime concept, for while a messier view of regimes makes for a more awkward and trickier science, it will reflect the world as it is, not an idealized version of it. The ultimate test on St. Marks is how well it works for its birds and woodlands, not how accommodating it is for scientists and administrators. As St. Marks shows, ideas about fire need to be as varied as fires.

EAST IS EAST, WEST IS WEST

Deseret Ranches

IN THE AMERICAN WEST grazing broke fire regimes. In the East it sustained them, and in Florida ranchers were largely responsible for keeping fire on the land as it industrialized and people poured into cities. Florida cowboys had burned in the spring, and not a few of them reburned large swathes in the fall. Add all the acres burned, and you could in principle get a number larger than the landmass of the state.[1]

Open-range herding, and the open-range burning that was inseparable from it, had been prominent features of the Florida scene from the first introduction of livestock. But it became a dominant feature after the loggers and turpentiners had stripped out the longleaf and then skipped out. Ranchers expanded into that void, plying the major resource left to locals until hustling real estate and herding tourists triumphed over wrangling ornery cattle. Florida's cattle frontier was spreading even as the western ranch frontier settled into domestication; the routine mayhem in places like Arcadia could make Dodge City look like Disney World. The Lincoln County wars were a feeble prelude to those in Collier County. In 1895 Frederic Remington visited Florida and considered its cowboys as mangy as their flea-bitten Texas ponies and their tick-infested cattle. Revealingly, Florida kept its range unfenced longer than any other state (until 1949). The man with 40 acres and 400 cattle became a staple of regional folklore, and so were his fires.[2]

But the need to eradicate Texas ticks and to keep mangy cows and bleary-eyed drivers from colliding on roads, the imperative for greater productivity in a rural economy barely past subsistence, and the realization that a brighter economic future lay in droving tourists and retirees rather than bovines led to a gradual shift in the character of ranching. What pine plantations for pulp did for forestry, a new generation of ranchers did for herding. The one constant for both was that they continued to burn.

This was the Florida context for Deseret Ranches. The larger setting was a determination by the Mormon church to invest in sustainable industries. Agriculture suited its temperament and seemed to embody the biblical principle of stewardship, an imperative to put the land to productive use and with a sense of accountability beyond crude ledgers. (A couple of years after it began purchasing the land, Ezra Taft Benson became Secretary of Agriculture in the Eisenhower administration, before ascending to the presidency of the Church of Jesus Christ of Latter-Day Saints.) A church that had emerged out of an agricultural frontier, one that thrived on pioneering, decided to expand into ranching outside the Mormon cultural hearth, the old state of Deseret. A colonizing impulse that had established agricultural outposts north to Canada and south in Mexico, now headed east to Florida.[3]

From 1949 to 1950 the church identified land in central Florida that could accommodate large-scale ranching. It bought the land and sought to develop it by the methods through which it had historically planted farming settlements throughout the Intermountain West. A group of families from Humboldt County, Nevada, was "called" to a mission to relocate the full apparatus of western ranch lore. The eastering pioneers not only brought the skills necessary to convert a rude ranching scene into a more substantive one, they brought in effect a small village.

The transplant failed to thrive. There was too much that was different, and there were too many families to support. Methods that worked in Nevada failed to root in Florida; like citrus, the new would have to be grafted onto the indigenous rootstock. Most of the families called were released. A decade of hybridization, overseen by a cattleman from

Arizona, Leo Ellsworth, followed, and when it ended, Deseret Ranch was flourishing as fully as its Brahma cattle hybrids. The ranch hired Florida cowboys, bought more lands, improved more pastures and more stock, and expanded into citrus, shell mining, and hunting leases. Although there is no coffee served in the office or over the campfire and business closes on Sunday, it is a commercial outpost, not a colony.

Today Deseret Ranch is the largest cow-calf operation in the United States. Although its cowboys still ride horses rather than ATVs, the 290,000-acre ranch coexists with power lines, three endangered species, and 45 hunting clubs. Viewed geographically, it sits between the urbanized tourist Space Coast and the urbanized theme park that is Disney World, and it represents a mightier version of the citrus and ranching industries that long defined the Florida interior. Viewed historically, it is a robust remnant of the ranch life that, ecologically, bridged the preindustrial landscape with preserved parks and public forests.

The critical facts are what ranching did and did not do. Ranchers did not eradicate or pave over the Florida biota, however broken and diminished from its prior condition; they kept the ecological pieces, even if they couldn't put them back together in working order. But they did burn. They had to. You couldn't ranch and not burn. Like subtropical grasses everywhere, the native wire grass was palatable and nutritious only when freshly growing, with a high protein content (27 percent) that would rapidly fall off as it matured. Raising cattle on indigenous range required constant fire to promote fodder and fight back the scrub. Ranchers burned as often as the landscape would carry fire, knowing that it would burn patchily (or "skip"), that some years were too wet or too dry, that cattle and flame competed for the same feedstock, that lightning and other people would kindle fires as well. At Deseret Ranch the standard rotation was three years.

Over the past few decades, transfixed by the wildland-urban interface, the American fire community has obsessed over borders generally. But Florida ranching offers another perspective. It is not about interfaces so much as interchanges. Pastoral burning over vast extents of the landscape—and because of open range laws not rescinded until after the World War II tourist and real estate boom was under way, the ranchers' reach far exceeded their grasp—those vast pastoral fires were a holding action that kept fire on the land. They massaged rather than changed the biota. By keeping fire, both their own and those set by nature and neighbors, they sustained at least fragments of the indigenous flora that had inspired Florida's name.

The scene did not change until the latter half of the 20th century. Paradoxically, the cattle frontier had expanded in Florida even as it was receding in the Far West. But eventually it, too, struck its boundaries and like a ripple in a barrel began to bounce back on itself. The need to eradicate the Texas tick (which required dipping each animal), the growth of fences and roads, and the increasing reach of formal fire protection (and concerns over smoke) all began to constrain burning onto lands actually owned by ranchers. Within those lands the changing economics of livestock argued for improved pasture and breeding, larger spreads, and a closer husbandry; the growth of roads and exurbs shut down the open landscape of free-ranging fire. Ranchers continued to burn with good effect, but they could burn less, and they no longer depended on burning alone. They fattened their stock on imported pasture grasses like Bahia rather than wire grass. They raised complex hybrids rather than rawboned cracker cows, and their fires showed the same evolution. Ranchers stood to make vastly more money by selling to developers than by leasing to hunting clubs and small operators. The squeeze was on, as it was for commercial timber plantations.

Ranchers remain a prominent feature of central Florida especially: they hold the land that will, depending on temptation and temperament, convert to nature parks or golf courses, stay rural or morph into retirement communities. For the American fire scene their greater significance lies in their historical role in holding fire on the land and in dramatizing, once again, the stark differences between the fire histories of east and west. The 100th meridian still marks a divide as profound for fire geography as the Mason-Dixon Line for politics.

The western story is one in which the removal of the indigenes and the introduction of livestock stripped away a major source of ignition and the fine fuels that fed fire. Over and over, throughout the semiarid West, the collapse of free-burning fire maps with uncanny fidelity onto the onslaught of sheep and cattle. Florida reverses this scenario. Cattlemen filled the demographic void left by disease and war—they topped off the fire potential not filled by lightning. By burning for forage, they also kept the grasses that would otherwise shade out under hardwoods, palmetto, and titi. They grounded burning in commerce, not simply tradition or a latter-day sense of ecological stewardship. Burning paid; not burning cost. It cost in lost pasture, and it cost in the expense of suppressing wildfire.

A real fire history for America is a federated one in which national and local stories each find a place and tumble into accommodations. Western notions of proper ranching didn't work in Florida, and had to be adapted; and that is pretty much what happened with fire institutions. An understanding of fire ecology based on the North Woods or the Northern Rockies, a conception of grazing and fire grounded in Arizona or Nevada, didn't work in Florida, and had to adjust, much as they had to deal with alligators instead of coyotes. The resulting hybrid has some of the best traits of each. Florida can fight fire with the best, while finding ways to discipline rural burning. Florida's engine and plow crews are interchangeable with those elsewhere in the country. Yet it has crossbred folk fire practices with modern technology to make more modern stock, as Premo-equipped helicopters replace cowboys dragging smoldering ropes across dry prairie.

Each region, each culture of fire, however tenacious its local pride and insistence that it is unique, implicitly assumes its own model is universal. Yet there are barriers to transport that make transfers difficult, and rightly so, like restrictions against the shipping of tick-fevered cattle. The usual history of settlement tells of a frontier leaping westward until the rainfall thins and eastern notions of farming founder in the arid West. Deseret Ranch, however, tells of western notions that stalled in the humid East; and that is what happened with fire practices and policies.

This simple symmetry of different settlements, east and west, may not end the story, however. The reverse may be no less true: there are few examples of the Florida model that have gone west without serious modifications, without local circumstances challenging and chipping away at the ideal. Florida cattle and open burning don't work in Nevada. So, too, it may prove with Florida notions of prescribed fire. However compelling the idea, however robust in the Florida flatwoods, the practice may have to hybridize in order to thrive elsewhere. Each region will have its own variant of landscape burning.

The real story of fire and ranching may not be one of concepts and techniques but of accommodation and the humility to seek out a practical reconciliation. Where the newcomers stubbornly hold to the old ways, they fail. Where they are willing to adapt, they can succeed. In the end, what Deseret Ranch successfully relocated was not a settlement, a colonization in the traditional understanding, but a sense of stewardship. The ranch is now a Florida institution, fully naturalized, although with some telling traits from its origin and continued ownership preserved and bred into the operation. There is a Latter-Day Saints ward house down the road at Deer Park that wouldn't be there had the Nevadans not been called. Senior missionaries give tours of the restored ranch house—snowbirds doing church service rather than spending days on a golf course or a recreation center. And there is, at least in principle, an official commitment to a sustainable working landscape.

That's not a bad model for what might happen with fire. Not every landscape can take the biennial blast of flame that historically renewed the Florida countryside: other places will need their own regimen. Few places can say that every day is a burn day: they will have their own seasonal and political boundaries. The Floridian with his promiscuous driptorch is as much an outsider as the Californian who dispatches an air tanker and a bulldozer to every smoke. In fact, when lawsuits have been filed for excessive backfiring on wildfires (with damage to private property), the agents have often been crews from the Southeast.[4]

So it's a good bet that as Florida fire practices head north and west, they will likely keep a few vestiges of their core legacy, not least a commitment to burning, but come to modify them to look a lot like their new surroundings. They will adapt and blend in; they will become naturalized, as most technologies and immigrants everywhere do.

A TALE OF TWO LANDSCAPES

Myakka River State Park and Babcock-Webb Wildlife Management Area

HIS CAREER WITH THE Army Corps of Topographic Engineers took Lieutenant Joseph Christmas Ives to two of the more intractable regions of the United States. From 1857 to 1858 he led an expedition up the Colorado River to a rumored Big Cañon. He believed the expedition would make him famous, and it did, but not least because he pronounced that his was the "first and would doubtless be the last party of whites to visit this profitless locality," what the world would come to know as the Grand Canyon. The army's concern lay with possible supply routes to the so-called Utah War. What is often forgotten is that the redoubtable lieutenant's previous assignment had taken him to the Big Prairie region that slashed across Florida north of Lake Okeechobee. It was "a comparatively unknown region" whose "natural features oppose great obstacles to the prosecution of surveys and explorations." In brief, the place flooded seasonally. In this case the army's interest lay in pursuing the Seminoles south.[1]

As transportation routes Ives's opinion of both sites was not far wrong. Their future, however, did not belong with passing armies and insurrectionists but with tourism. For the Canyon the scene was primarily geological. For the Big Prairie it was biological. Eventually the Grand Canyon became the largest unit in the national park system. A patch of Big Prairie became the flagship unit of the Florida Park Service (FPS).

Historically, the region around the Myakka River boasted a typical Florida landscape, seasonally overflowing either with water or flame. The scene came with one enormous geographic anomaly, however, a large swathe of savanna pocked with wetlands, pine "islands" and oak-palm hammocks and seasonally washed by flame, what was known as the Big Prairie and later as the Florida dry prairie. The area seemed so barren that early surveyors skimped on their markers in the belief that no one other than migratory hunters and ranchers would find enough value to visit the place routinely, and no one was likely to homestead. Unsurprisingly, settlement was slow, as it was throughout interior Florida. A few low-grade railways supported high-grade longleaf logging and scabby naval stores; the woods were too scattered and sparse for serious milling. Staked ruts for wagon trails marked routes of travel. Mostly, cattlemen ran cows and kept the landscape low and open through annual burning. By 1930 fewer than 39,000 people occupied the region now comprising Manatee, Sarasota, and Charlotte Counties, and nearly all resided on the coast. Lightning, transients, and ranchers burned most of the land on an annual or biennial rhythm.[2]

The big change came in 1934 when the interests of the Sarasota Fish and Game Association merged with a New Deal that was busy scouting sites for Civilian Conservation Corps camps. The land acquired to support Camp SP4 became the nucleus for Myakka River State Forest and Myakka River State Park, which in turn evolved into the flagship site for Florida's protected state lands. The enrollees promptly did what they did everywhere: they built infrastructure, planted trees, and fought fires. Since the biotic core was prairie, not forest, and the biological dynamics relied on routine burning and flooding, the CCC inaugurated an era of ecological irony because they broke both flood and flame. Among the infrastructure they bequeathed, along with roads, weirs, drainage canals, and dikes, was a policy of fire exclusion outfitted with hundreds of miles of firebreaks and hundreds of flapper-wielding enrollees. They stopped deliberate burning, and they tried to stop nature's fires as well. In this the fledgling Florida Park Service, initially overseen by the state Forest Service, embraced the collective wisdom of nature preservationists everywhere.

Outside the Myakka River State Park the land continued to burn—ranchers saw to that; and what they missed, lightning scavenged. Other conservation agencies acquired sites nearby. In 1941 some 19,200 acres were bought from rancher Fred Babcock to become the Charlotte County Preserve, which over the next 40 years grew into the 65,675-acre Babcock-Webb Wildlife Management Area (WMA) administered by the Florida Wildlife Commission. Interested mostly in fostering further hunting, the Wildlife Commission encouraged ranchers to burn under a regimen of grazing leases. In 1962 the site hired a biologist, who added his own reasons for burning. As money for land acquisition kicked in during the 1970s, and especially the 1980s, more parcels around the region transferred from the private to the public sector into what enthusiasts began (hopefully) referring to as the Myakka Island, the whole of the watershed linked by wildlife corridors as well as by streams. Except where converted to houses and malls, most of the new lands persisted in their old fire practices. Paradoxically, the most ostensibly protected places, the parks, were the exception, and Myakka River, which became a park in 1941, was, at enormous cost, protected from the flooding and firing that had made it flourish.[3]

The Florida Park Service divorced from the Forest Service in 1948. The split made sense since parks and forests had different purposes, but it left park management without the economies of scale—intellectual as well as material—that a larger institution offered, and it left park administration without a conduit for resource management rather than a program of strict protection. By 1969, as Florida was voting in a new constitution, the FPS readied itself for a major overhaul of its own. The reformation sent its new chief naturalist, Jim Stevenson, to Tall Timbers Research Station to see for himself what the brouhaha was about. At the time the FPS was the only public land agency in Florida that did not prescribe burn. Stevenson recalled, "We permitted no prescribed fires on state parks. Fire was just the worst possible thing that could happen, and we would risk life and limb to put out any fire that occurred."[4]

Ed Komarek gave the naturalist a field tutorial. ("I believe Ed Komarek could sell a forest fire to Smokey Bear," Stevenson later wrote.) He left convinced that the FPS had to burn. Shortly afterwards Betty Komarek conducted the first prescribed fire on a state park (Falling Waters). By then 20 years had passed since the Park Service had achieved autonomy and spun off from the emerging fire program of Florida agencies. Over the next 20 years it tried to make up for lost time and sought to put fire in as vigorously as it had previously struggled to take it out. In 1989 Jim Stevenson informed a Tall Timbers fire ecology conference that over the past year the agency had burned 110,000 acres in 66 parks.[5]

The agency seemed to be holding its own. But by the time Stevenson had gone to Tall Timbers in 1969, 35 years of attempted fire exclusion had already passed. Myakka River's ledger of fire removed was, in reality, the record of a crushing fire debt.

From the vantage point of the present, the tale of two "natural areas"—Myakka River State Park and Babcock-Webb WMA—can make painful reading. The one compensation, perhaps, is the old adage that great heroes require great villains. The more menacing the challenge, the more admirable the achievement. The contrast in fire histories is stark.

The more rudely managed site, the WMA, had never abandoned burning. From the time it was gazetted, cattlemen had continued to burn under leases, and when an official biologist arrived in 1962 he tweaked the regimen but left it intact. Unlike the Florida Forest Service, the preserve did not burn primarily for fuel reduction as leverage to help control wildfire but for habitat suitable to cattle, deer, and quail.

For the first 20 years the fire regime was classic Floridian: it free-ranged as much as the cattle. The legendary Buck Mann described the routine on his ranch just south of the park. Every year he drove his Jeep over the dirt ruts that passed for a trail, and whenever he spied a patch that needed burning he struck and tossed out kitchen matches. The fire would "burn wherever the wind carried it. It would burn until it reached a wet swamp, marsh, or hammock. Sometimes the fire burned for days." There were no control lines, no burn plans, no prescriptions, no backup crews for suppression, no firing strategies to lessen smoke, no standards for burners, no radios, no Nomex clothing or hardhats. There was no accommodation for species—the cattle moved out of the way, the other creatures thrived on the postburn flush. The dry prairie revived from its seasonal fire drought. There was no particular attention to neighbors, who were all ranchers; flames blew past fence lines as easily as over gopher holes. If Buck hadn't burned his neighbors' lands, they would have burned his.[6]

For the next 20 years, as the WMA expanded, fires remained long burning, tacking and veering with storm and wind, which meant they were large. As hunters relied more on vehicles (which made roads) and as population began to swell along the coast, managers gridded the preserve into 160- to 240-acre blocks etched by plow lines, which made control easier and long-lingering smoke less annoying to neighbors and less hazardous to I-75, which cut through the site. For the next 15 years the fire rotation flickered around two to three years. For the last 15, up to 2010, it ratcheted down to one to two. Essentially the land burned as often as it could.[7]

The contrasts with the iconic Myakka River State Park are stunning. The blessings Myakka River enjoyed at its birth turned into banes. The Park Service was statutorily charged with preserving "representative portions" of "original natural Florida," which it interpreted as overthrowing the inherited rural regime. It excluded the cattle and the hunters, substituted tourists for traditional transients, and banned any burning. The CCC gave it the clout to physically attempt to replumb and replant the landscape and to fight fires. Enrollees platted the landscape with firelines. Not until Stevenson had his epiphany and the FPS saw that it could not restore cherished landscapes without restoring fire did managers again pick up the torch.

Meanwhile, the Big Prairie was mutating into the Big Palmetto Patch, and the Florida grasshopper sparrow, burrowing owl, Sherman's fox squirrel, and crested caracara were gone. A sun-loving flora was being literally overshadowed, and a fauna of spotted skunks and gopher tortoises that depended on them was becoming rare. The "original domain" was choking on an obese, rough, and thickening saw palmetto, and ecologically suffering as if from biotic diabetes. The dry prairie had become the fire analogue of a drained wetland that was rewatered only by damaging floods. So it was with fire: savage wildfires broke out, and suppression replaced nourishing ash with plow lines.

But in a place where, in the absence of cleansing burns, palmetto can shoot up to head-high thickets, shrubs can often overtop grasses and forbs in less than a dozen years, and pine islands thicken into backlot woodpiles, all capable of stoking conflagrations after as little as five years, bad fires promised to replace good ones. Bad fire was the original ecological invasive: these fires did not burn as they should, much as Norway rats did not behave as possums. Myakka learned that putting fire back was more formidable than taking it out. After being excluded for more than three decades, probably fire acting alone was not enough to restore itself. Reinstating something like fire's old regime was bound to be tricky, and often crude, and occasionally violent. The earliest efforts were all of the above.

The project began with efforts to understand what the conditions were at the time the park was established. Written accounts, original survey notes and maps, aerial photographs, interviews with old-timers, all pointed to an open savanna with few trees, most of them clustered into pine islands, and Myakka as a place that burned relentlessly and benignly. The first prescribed burns followed the formulas of the day; winter fires, backing fires, tightly leashed controlled fires held within hard lines. But fire is what its context makes it: fires in a beaten biota behave like trodden-down burns. They could barely hold against the continued pressure to biotically bury the prairie and overgrow the pines. They might maintain the existing scene but could not reverse the trends, much less restore the old. They would have to reverse drainage by filling in canals, fight off exotic plants that burned on different rhythms, and probably crush the rough. The Myakka fire staff realized, too, that the park no longer received fires from adjacent lands, that roads and boundaries had grown into unburned hedgerows. Fire's habitat was as fragmented as those of caracara and gopher tortoise.[8]

In 1985 the staff adopted more aggressive measures, and a feistier mix of fires. They experimented with brush cutting and roller chopping to bring some mechanical leverage against robustly rooted ligno-tubers, intrusive woods, and the ever-more-entrenched palmetto. They loosed heading fires that might bring some thermal leverage as well, hoping a blast of heat could shock the system in ways that slow-cooking backing fires could not. They kindled growing-season fires in an effort to reintroduce a more natural regime. They watched fires hop and skip across freckled landscapes. They accepted some harsh burns and scorch-killed trees as the cost of paying down the park's fire debt. Yet even as the program's tempo quickened, countervailing frictions threatened to drag it to a halt.

The 1985 season was a horror across the state. On May 16 a wildfire brought much of Flagler County to its knees and incinerated 131 homes at Palm Coast. At almost the same time, what Robert Dye called "a letter based on a second hand description of a successfully conducted burn, claimed that 'it was by the grace of God my horses were not killed and property destroyed,' was addressed to the Governor." The governor passed the buck to the Department of Natural Resources, which proclaimed a moratorium on all burning in state parks through the summer season. Later, Myakka River endured its first off-site escape fire (onto an adjacent ranch), which kindled another ban, this one specific to the park.[9]

Yet, even as it became harder to burn, the necessity for still more burning intensified. Without tame fire, feral fire would overrun the park like a pyric version of cogongrass. By now resident species were being federally listed and the Florida dry prairie itself was recognized globally as a habitat at risk. Myakka River State Park was in danger of reclaiming its iconic status but as an emblem of unintended consequences. In the short run, a careerist knew it was safer to do nothing than to do something. But Myakka River was blessed with managers like Dye and Paula Benshoff, who remained on the scene for their whole careers. They were willing to bet on the long run, and that meant getting control over the fire regime by getting fire back on the land on their terms. The park received more equipment, cut broader perimeter fire lanes, and reinvigorated its experimentation. No one was fully happy with the amount burned, but they were no longer falling behind. Myakka became an exemplar of fire management for the park system overall.[10]

By the 1990s the prairie was being slashed and burned—chopped by tractor-drawn rollers and burned with a diversity of line fires, spot fires, heading and backing fires, dormant and growing-season fires. Fire managers aimed for a two-year rotation on restored lands and continued conversion of lost lands to a condition that better approximated the "original natural" state. But nature doesn't operate on a nightly news cycle or bureaucratic budgets. That revanchist rough could not be stockpiled in caverns like inert nuclear waste: it grew and it was ever ready to burn. It was seemingly impossible to dislodge palmetto once it sank roots and thickened into swathes. A missed fire or two might allow the rough to reclaim what exhaustive labor had restored, and a handful of years without the right kind of fire is enough to lose what decades of onerous labor had clawed back. While students of public administration might speak loosely of "institutional ecology" or some other kindred metaphor, the

Florida Park Service ran on energy flows and nutrient cycles very different from those that operated on the Florida dry prairie. Probably, too, part of what was missing was the biological interactions that fire had catalyzed, and now no longer worked—a dynamic version of fragmentation. By the onset of the 21st century the park had devoted as much time to restoring fire as it had previously invested in excluding it, yet a backlog of burning still brooded over the scene.

More than nature pressed back. There were social pressures not to incinerate a favored recreational site or to immerse a booming Sarasota County with smoke. Particularly, granted the amount of migration into the region, public education required a continual program of messaging to keep the rough of public opinion at bay. As the Myakka Island concept of a regional consortium of linked lands devoted to conservation gathered momentum, it added to the stockpile of mandatory burning, but it was harder to get money for maintenance than for acquisition. No one's name went onto a plaque honoring their investment in biennial burning. No one would be memorialized for saving fire as they might be honored for saving black bears.

Nor did all the restrictions come from the outside. Bureaucracy imposed an increasing roster of restraints. Burning required more than just plans, prescriptions for flame and smoke, checklists for equipment and crew size, and Florida Forest Service authorization. Crews had to pass training and fitness standards. The Keetch-Byram drought index established the statewide conditions for determining whether a particular site could burn or not. Individual parks had to submit to agencywide policies such that a failure in a panhandle park might affect how Myakka did business. The restraints were cumbersome, but necessary; they served as social buffer zones, the legal equivalent of perimeter fuelbreaks. Still, the program that had reinstated fire from the 1970s onward was a far cry from Buck Mann tossing kitchen matches from a Jeep as his father had from horseback. But the fire managers of the early years well into the late 1990s enjoyed a freedom of action that the next generation would never know.

The dilemma characterized the entire system, however. The need for fire exploded, while the capacity to burn did not keep pace. Over the Florida parks system as a whole, the anxiety thickened that fire managers would be able, with ever more strenuous effort, to hold the line, but little more. In 2008 the FPS burned 89,000 acres on 66 units, although critics noted that four parks accounted for 80 percent of the total and that it appeared that some places might be simply blackening acres, not burning for identifiable ecological goals. The fretters worried that, with respect to fire, they could make the monthly payment but might not be able to pay down the accumulated debt.

Still, 40 years after Jim Stevenson had made the pilgrimage to Tall Timbers and got religion, the Florida Park Service had successfully gone from a program of fire exclusion to one that burned its "fire-type" holdings on a roughly four-year cycle. A third of its units met their minimum targets. Wildfire accounted for only 2.4 percent of acres burned. The agency began to track acres burned for maintenance apart from those fired for restoration.[11]

The state parks' story is a cameo of Florida fire generally. Most land continued to burn routinely (not always wisely or well) until it fell under formal protection. The Florida Forest Service estimated that a third to a half of those lands not legally protected burned annually. Then came the great transition. The postwar boom transferred land from rural status to urban or public, and with that shift came fire exclusion or, out of the attempt at exclusion, a remaking of flame from tame to feral. In many places, the continuity of fire broke, as it did at Myakka River. In others, it persisted, though altered, as at Babcock-Webb WMA. The modern saga of Florida fire hinges on that transition.

The Florida Forest Service accepted prescribed burning at the same time legislation closed the open range. In principle it was possible to substitute forestry's hand on the torch for ranching's; but the free-range fires of the cattlemen far outstripped the more tightly controlled burns of state-sponsored forestry, which were intended not to spring-clean landscapes but to batten down fuels as an aid to controlling wildfires. By 1970 formal protection extended throughout the state. Between the closing of the range and the saturation of fire protection lay 20 years in which burning lagged, and during which, for the Florida Park Service, only wildfire was possible.

The astonishing land-acquisition programs quickened the move from rural landscapes to preserved ones. Again, burning lagged, as administrators waited to install infrastructure, train and equip crews, sponsor research on burn plans and prescriptions. Even when the torch passed, a step or two was lost in the handover. In the case of small units, the burning never happened or occurred only in token daubs. In others it came too late, and fire had to be restored. As with a lost species, lost fire first required a suitable habitat if it was to flourish. Nature, however, didn't wait for the bureaucratic planets to come into alignment. It sprouted fuels. A handful of years was enough to mutate the fire regime from annual surface burns that brushed and skipped over the landscape into spasms that incinerated drought-readied rough and slammed against suburban sprawl. The old regime could not wait 40 years, or 20, or even 10 for a new one to seamlessly replace it. Five or six years without fire meant that burning was no longer a casual task in an afternoon drive but a costly and potentially onerous bureaucratic exercise. The legacy of even a decade meant a debt of lost fire that might never be made up. Few places burn as often and widely as managers believe necessary.

Paradoxically, those pieces of protected nature that came later were often better disposed to fire than those that had come early. Places that had endured a fire-free phase would have to be biologically rebuilt in order for fire to do its ecological work, and it was not obvious that fire alone could remake the old setting. Flame remained, as ever, an interactive technology.

Myakka River State Park demonstrates the old adage that it is easier to tear down than to build up. The oddity is that, with Florida fire, doing nothing could be the same as tearing down. Not burning might do as much harm as clear-cutting or overhunting; reinstating fire was far trickier than replanting pines or creating nesting trees for osprey.

It was not enough to cease suppressing fires. They had to be deliberately restored, and since fire and land were entangled, that meant that both fire and land would have to coevolve into whatever condition people sought. It wasn't enough to plant, slash, or uproot exotics without burning, but neither was it sufficient to dump fire on the land. It was a process in which each interplayed against the other, while people constantly relearned from each new outcome. During his conversion experience, Jim Stevenson recalled that Ed Komarek had noted that it had taken 30 years of mismanagement to trash the parks, and it would take 30 years of "proper management to restore them." That seemed a long time, even if the perspective offered a pleasing symmetry. It also underestimated the time required.[12]

How to restore fire has proved a far more vexing task than how to suppress it ever was. Yet a gamut of strategies has appeared that ranges between coaxing and coercing. A Florida advantage is that the sites mostly remain working landscapes, so there is no ideological block against manipulating them by using chemicals, chainsaws, masticators, and tractors. The western parks, dominated by wilderness concerns, have fewer options. That's one distinction, east and west. The other is the mix of urgency and patience that fire managers bring to the task. In the Far West decades may pass before the land has so altered that putting fire back is complicated or ruinous. In Florida one has only a handful of years before returning fire means restoring it. Oddly, it is the westerners, however, who are keenest to make the revolution happen instantly. The southerners have a greater forbearance toward the long rhythms of history.

No theory dominates practice. Restoring fire—nature's agent of creative destruction—to nature's economy is not unlike restoring capitalism to an economy stagnated by decades of state communism. One strategy is shock therapy: dump it on the system and let the system sort out the mess. The informing belief is that fire is inevitable, the landscape can absorb the blow, and the sooner it happens the better since it allows renewal quickly. To wait is to worsen. Especially in natural settings, the invisible hand of nature will guide a rapid recovery.

The Florida strategy favors coaxing. It takes a longer view, expressed in a more incremental approach; landscapes segue one into another. Ideally, restoration proceeds bit by bit, as the undesirable parts of the landscape are removed, and with fire as catalyst, the desired parts revive. Retaining canopy trees, for example, even if they are the "wrong" conifers, keeps fine fuels on the ground through needle cast, which allows fire to persist. Likewise, keeping old burning practices, although they are not precisely targeted, holds fire on the scene. Once broken, fire and fuel, unlike bones, do not rebond stronger than before. Throughout, the hand of management is visible.

As the Myakka story shows, coaxing does not mean dawdling. The park does not have the luxury of reassembling the landscape piece by piece in cautious sequence. It's rather an exercise in coprocessing. It embodies what Horace once expressed as *festina lente*—hasten slowly. The prepark generation burned under circumstances that seem today unspeakably lax; the generation that restored fire after three decades of abusive exclusion operated under tighter constraints, both socially and ecologically, and confronted a crushing backlog of bad ecological debts. The generation that succeeds them will face an environment more tightly coiled still, requiring additional effort simply to hold existing levels of burning. Instead of uprooting planted slash pine, filling in ditches, and loosing fires, they must cope with such ecological game changers as Brazilian pepper trees, West Indian marshgrass, Cuban tree frogs, tighter regulations on particulates in smoke, closer suburbs, landscapes broken up and spliced together without regard to fire. They will have to burn both more and less, both faster and hotter, yet all with finer precision. Using flame they will have to beat back the unwanted rough, while, using ideas, they will have to fight off the ceaseless challenges to burning that crowd the margins like invasive weeds. Throughout, they will need the will to burn, which may ultimately depend on what William James once termed "the will to believe."

Today the distinction between maintaining and restoring still has meaning. In the future the two concepts may merge. It may be that Florida's experience with fire restored is an artifact of a particular historic time in which a remarkable transition in land use demanded a set of fire practices different from those that had previously kept fire on the land. That experience seemed, at the time, a unique moment. But as pressures mount, it may become the new norm. It may be that the only way to maintain fire is to continually restore it.

ONE FOOT IN THE BLACK

The Nature Conservancy in Florida

IN THE AGE BEFORE Standard Orders and rosters of Watch Outs, safety training relied on a few bits of folk admonition. To raw crews: "Stay with your foreman. He has been to many fires and is still alive." Against a crown fire: "Pray for rain and run like hell." And for fast-paced hotline: "Keep one foot in the black," which recognized that the safest place to be if a fire roused to fury was in its already-burned patches.

It's a wonderful phrase, though, and with a bit of metaphoric tweaking it might stand equally for the admonition to a landowner in a fire-prone place that working the land means working with or against a line of fire and that controlled burning is preferable to wild burning. Land management is not planning: it is doing. If you are serious about running a place that holds or needs fire, you have to work with one foot always in the black.

That line—between the black and the green—might equally divide nature organizations. There are the greens that study and plan and exhort, and there are those that must also work the land, that straddle an advancing line of flame and must keep one foot in the black. In fact, that's not a bad way to characterize the Nature Conservancy (TNC).[1]

The institution evolved out of a desire by a group of ecologists for "direct action" to protect natural areas. In 1951 the ambition incorporated as nonprofit organization. The Nature Conservancy created a network of chapters, beginning in New York, that eventually spread throughout the United States. Over the next decade it established its trademark strategies: land acquisition, conservation easements, partnerships with public agencies, and a reliance on local volunteers. What it bought it had to manage, although in some cases only until the land could be sold to a public agency. Most of its initial purchases were small—60 acres of the Mianus River Gorge, six acres of the Bantam River marsh. Not until 1965, with funds from the Ford Foundation, did TNC have a full-time president.

What drove the Conservancy into fire was its acquisition of prairie in the upper Midwest. The only way to sustain prairie was to burn it. In 1962 it conducted its first burn on Helen Allison Savanna Preserve outside Minneapolis. It was done like the rest of its land stewardship by local volunteers. Four years later, however, Katharine Ordway, heir to the 3M company fortune, began donating money to buy tallgrass prairie, first in Minnesota, and then throughout the Great Plains, on what seemed an industrial scale. As it burned more, the Conservancy, which had no internal fire culture, came into contact with places that did. In Wisconsin it met an academic clique interested in restoration and accustomed to work on small plots of oak prairie savanna. As it moved into the Flint Hills of the central plains it encountered a stubborn culture of burning that dated back to settlement, when farmers found the hills too rocky for plows and abandoned it to ranchers, who found they had to burn to boost forage and protect against wildfires. The Nature Conservancy became, by default, a fire agency. By 1985 its involvement had advanced sufficiently for Mark Heitlinger to write a general manual for fire operations.[2]

What made TNC different from most environmental advocacy groups was that it owned its lands. What made it different as a fire agency was that it emphasized burning for ecological goals. What put the program on steroids, however, was land acquisition in the southeast, particularly Florida. No landscape burning is simple or inherently safe (people even lose campfires all the time). But tallgrass prairie is relatively homogeneous; burning the same sites over and over builds experience that leads to competence and confidence; and most larger prairie sites abut other lands of similar composition that can accommodate some spillover flames. Little of that applies to Florida. The Conservancy had to burn on a scale and under conditions that exceeded what it could expect from local volunteers, and Florida marinated the organization in a regional culture of prescribed fire very different from that of the Midwest. The southeast office decided it needed a dedicated fire staffer. In 1986 it hired Ron Myers, a University of Florida PhD with fire experience who was then working at the Archbold Biological Station.

If its Florida experience was the pivot point for TNC as a national fire organization, and eventually an international one as well, Ron Myers was the bearing on which that transition turned.

His career is in some way a cameo of the national story. He grew up in one fire culture and converted into another. A Californian, he spent a summer on a timber stand improvement crew on the Shasta-Trinity National Forest in 1965; he got on a few fires and acquired a yearning for more. The next year he enrolled in forestry at the University of Montana and worked that summer as a fire control aid at Glacier National Park. He was at Glacier in 1967 when the Flathead and Glacier Wall fires broke out, and spent a month on firelines. The effort to suppress those fires, what appalled many critics as mechanized brutality, helped push the National Park Service into its policy of fire by prescription. The next year was empty of fires. In 1969 he joined the Missoula smokejumpers for a season full of them. He learned fire by fighting it in the settings that, since 1910, had most shaped national policy.

Then he commenced a long reeducation. In 1970 he enrolled in the Peace Corps and for two years taught at the National Forestry School at Siquatepeque, Honduras. That brought him into contact with a living culture of rural burning in many ways alien equally to smokejumping and to the formal learning he promulgated through lectures. He was astounded to find pine forests that burned routinely, deliberately fired to good effect, though sometimes to bad. When his tour ended, he signed on as a seasonal at Everglades National Park, then making the great transition to full-spectrum fire management under Larry Bancroft. He yearned to return to Latin America and decided he would need an advanced degree to do it. Education was a means to get into the field.

He enrolled at the University of Florida. Its forestry program targeted pine plantations and had little passion for international work. He took up botany, studied under Jack Ewel, did field research at what became Big Cypress National Preserve, which educated him into the techniques of field biology, and as a teaching assistant developed ecology courses for nonmajors, which taught him how to translate complicated ideas for a wide audience. For his doctoral research he returned to Mesoamerica under an Organization of American States fellowship to study Tortuguero National Park in Costa Rica from 1977 to 1980, a sodden rainforest without fire. He returned to UF as a postdoc, during which he coedited with Ewel and contributed chapters to a book, *Ecosystems of Florida*—still the standard reference. From 1982 to 1986 he worked as an assistant research biologist at Archbold Biological Station on Florida's Lake Wales Ridge. There, he developed the station's first fire management plans. In 1986 he signed on as the Nature Conservancy's first fire staffer. In 1988, the year Yellowstone National Park scorched TV screens for a long summer, he became its National Director of Fire Management and Research. He anchored the program in Florida at Tall Timbers Research Station.

He hired Paula Seamon, and together—in Seamon's words—they "grew" a national program. Myers proved indefatigable. Since TNC did not have a resident fire ecologist, he had to evaluate all its sites, write fire plans, and train local workers, with extensive mentoring chapter by chapter. All those sites known to need fire. All those that needed fire but didn't know it. And not only TNC holdings but all those lands and partners with whom the Conservancy had agreements, had arranged easements, or undertook mutual operations. These became a serious commitment as TNC expanded rapidly. It had completed a Natural Heritage Network for all 50 states in 1974, launched an International Conservation Program in 1980, and the same year Myers assumed the directorship for fire management, signed an agreement to help manage the 25 million acres held by the Department of Defense. A year later TNC commenced its Parks in Peril program, targeting Central and South America and the Caribbean, and acquired the Barnard Ranch that became Tallgrass Prairie Preserve. In 1991 it inaugurated its Last Great Places initiative.

Fire management was integral to it all. At the onset of the new millennium, TNC had a dedicated fire staff of three—Myers, Seamon, and Jeff Hardesty—to oversee the lot.

The challenges were enormous. Its fire staff had to plan, train, and leverage. The tiny group had to transform conviction and determination into programs that could meet not only the astonishingly disparate needs of TNC but that would allow its stewards and volunteers to participate with partners. Each response rippled through the organization and, with surprising quickness, through the national and even international fire establishment.

First, there were the internal needs of the Nature Conservancy. It had to manage fire on prairies, scrub pine, barrens, and wetlands to promote endangered species and eradicate noxious invasives and do it all on a shoestring. To restore prairie and free-range bison, Tallgrass Prairie burned on a vast scale but through grasses, amid a ranching landscape that also burned, and within a legal environment of strict liability. To sustain habitat for the endangered Florida scrub jay at Tiger Creek Preserve along Florida's Lake Wales Ridge, a prescribed fire team needed to set high-intensity fires in tightly bounded patches, often with collaborating agencies, within a regulatory environment that required Florida Forest Service authorization. To fashion habitat favored by the Karner blue butterfly, Albany Pine Bush, which was managed by a consortium of six landowners, had to burn among discontinuous patches nestled awkwardly within Amtrak lines, two interstate highways, assorted state and county roads, the city of Albany, major electrical transmission lines, a landfill, three nursing homes, assorted private lands and houses, a police-fire state complex, and a trailer park with a cluster of propane tanks.

But the Conservancy also had to reconcile two distinctive fire cultures. One culture was a tradition of prairie burning maintained by ranchers. Its core lay in the Flint Hills, with flakes split off across the plains. While chapters undertook similar burning with the earnestness that seemed endemic to TNC enterprises, the practice had a quality of studied relaxation, not leisurely but measured, like the rolling hills and homogeneous grasses in which it happened. The Conservancy absorbed the other tradition from its full-immersion baptism into Florida's fire scene. Burning here was concentrated, more and more bounded, potentially violent and unforgiving. It was a domain that favored toughened cadres for whom burning was something done year-round, not part of an annual ritual of spring cleaning. No one would dispatch chapter volunteers from the plains for fireline duty off site. Florida's TNC crews could join any brigade anywhere. Between those polarities lay a scatter diagram of sites, through which it seemed impossible to draw any line of administrative correlation.

Ron Myers, however, argued that this gaggle had to be yoked to a national standard. The complexity of island ecologies made burning more technical, liability from a burn gone bad could ruin the Conservancy, and to share operations with state and national partners TNC needed to accept common norms for training and qualifications. Unlike government entities the Conservancy had to buy insurance on the open market, which meant meeting social norms of accepted practice to avoid charges of negligence when escapes happened. If TNC was to burn in Florida and receive liability protection, it had to satisfy state certification for burners, and if it was to join programs on federal lands, it had to satisfy national criteria. That pushed the Conservancy to adopt National Wildfire Coordinating Group (NWCG) standards, much to the consternation of many chapters.

The move gave TNC a place at the table and the opportunity to influence the character of such training, particularly with regard to prescribed fire. It could inject objectives for ecological burning into manuals that otherwise boxed in burning to fuel reduction. It could validate a fire organization based on prescribed fire rather than suppression. In fact, the national courses that emerged under NWCG evolved out of TNC material. Public agencies even sent their staff for TNC training. The Conservancy was doing what it did best: it built capacity, leveraged small resources into wide results, and served as a catalyst for making ideas operational.

The Florida program sprang out of the scrub-pine sandhills of the Lake Wales Ridge, as tough a landscape to burn as any in the country, and one ratcheted more tightly as 11 agencies oversaw 63 protected sites even as urbanization spilled south from Orlando. In 1991 the Florida chapter helped organize a Lake Wales Ridge Ecosystem Working Group (Ron Myers had come to the Conservancy from Archbold Biological Station at the south of the Ridge). In 1992 Disney Corporation created the Disney Wilderness Preserve at Kissimmee and funded TNC handsomely to run it for 20 years; the 12,000-acre project brought modern equipment and the imperative for smartly trained staff; and in response the Conservancy hired mostly out of the Florida Park Service. Among the fire community it became a valued partner and a neutral enabler. When, that same year, the controversy over seasonality of burning was peaking, the Florida Game and Fresh Water Fish Commission sponsored Ron Myers and

Louise Robbins, also with the Conservancy, to write a summary review and make recommendations. Full integration, however, came with the state's 1998 trial by fire. TNC stepped up, joined the suppression effort, and contributed several strike teams, which proved particularly valuable for nighttime burnout operations. Because it met standards, TNC could participate, and afterwards the Florida Forest Service contracted with the Conservancy to train the National Guard to meet them also.

The next stage in its Florida evolution was to formalize those firing teams. In 1999, as the Conservancy's fire program was bidding for national standing, it conceived a Florida Scrub-Jay Fire Strike Team (later renamed the Lake Wales Ridge Prescribed Fire Team) as a dedicated body to assist that consortium. The team had six members, two of them rated as Burn Boss II, along with three Type 6 engines and two ATVs. The team proved particularly valuable in grappling with the burning backlog problem, as a vanguard to conduct those first, tricky burns in overgrown settings. It soon became a vehicle for training and exporting to other TNC chapters; the Florida Park Service replicated the idea with a Backlog Abatement Team; similar consortia congealed for northeast and central Florida; and the experience was passed along to the Maine Forest Service and Army National Guard, the U.S. Forest Service in Colorado, and the Bureau of Land Management in Alaska, and then, as the fire initiative went global, so did the expertise of the Lake Wales Ridge Prescribed Fire Team.[3]

What happened in Florida, in brief, happened nationally. In the late 1990s the U.S. Forest Service seconded a staffer to TNC for a year. That experience led to a Fire Roundtable held at Flagstaff, Arizona, in 2000 to explore collaborative interests, an idea that then became embedded in the National Fire Plan. A presentation at the National Interagency Fire Center led to a general cooperative agreement with all the federal land agencies. Amid an increasingly polarized (and paralyzed) national establishment the Conservancy was viewed as an honest broker, a facilitator. The outcome was the Fire Learning Network, a spectrum of national training, and active partnership with LANDFIRE (Landscape Fire and Resource Management Tools). In 2006 the concept was replaced by a five-year "Fire, Landscapes, and People" program that continued and expanded the agenda. TNC seemed to be everywhere, and everywhere a van der Waals force that helped hold factious alliances together, and everywhere welcome. All this was in addition to TNC's own swelling population of sites; in some years the Conservancy prescribe burned more acres than the National Park Service. And it was complemented by a stunning Global Fire Initiative.

The Conservancy's international program picked up steam just when Ron Myers assumed the directorship of its national program. It was, again, an ideal collusion of institutional needs, c.v., and personality; TNC's national fire director was the perfect point man for its international campaign. With 80 percent of TNC global efforts directed to Central and South America and the Caribbean, Myers could draw on his personal past in Honduras, Costa Rica, and Puerto Rico to craft practical plans and build local capacity. Over the next 15 years he designed and sparked to life fire management programs and the training to make them happen in Mexico, Belize, Guatemala, Honduras, Costa Rica, Cuba, the Bahamas, Trinidad, the Dominican Republic, Peru, Paraguay, and Argentina, as well as short assignments for training and reconnaissance to China, Mongolia, South Africa, and Indonesia. In 2006 Myers codified the experience in a glossy-page publication, *Living with Fire—Sustaining Ecosystems and Livelihoods Through Integrated Fire Management*, a shockingly readable document in which the jargon of bureaucracy and science melted away in favor of direct language, categories that could hold a driptorch, a vision of fire ecology that included people, and a universal "framework" of fire management that quietly dumped the 10 Standard Fire Fighting Orders in favor of 10 "guiding approaches."[4]

The 28-page document was a personal testimony of a singular life, the synthesis of a Montana smokejumper turned world-class prescribe burner, a botany PhD who learned from campesinos in Honduras and Mexico, and a pragmatist who knew that ideas only had meaning as they were expressed on the ground. In many ways, too, it was the TNC model translated and exported: networks of small chapters trained to basic standards of safety and effectiveness while grounded in and sustained by local communities, to which outsiders might offer guidance but would not administer. The contrast between lumbering, costly public agencies in the United States struggling to meet prescribed fire targets and a Conservancy rapidly scaling up its programs on a shoestring with a permanent staff that could barely fill a Type 6 engine repeated itself overseas. In countries like Mexico it provided the leverage to allow local interests and institutions to take root.

Then the Great Recession took the wind out of TNC's sails. Like a body in shock that withdraws blood from the extremities, the Conservancy laid off a fifth of its staff, pulled back from international commitments, and shuttered its Global Fire Initiative. After 23 years of service Ron Myers was given notice.

The Nature Conservancy fire program was so small in its formal constitution—bureaucracy seems too strong a term—that it reflects to an unusual degree the places that nurtured it and the personalities of its officers. Its design favored the local over the large: that's how it was able to scale out rather than up.

Much of the TNC experience, in fact, had wider significance. The Conservancy was not tied to the eccentricities of particular landscapes, as the Tall Timbers model was. Nor was it pushed and pulled by legislative mandates, like the Florida Forest Service. Nor was every decision potentially a federal case, as with the national forests and parks. It demonstrated that one could burn for biodiversity as well as for fuel reduction. It showed that a vibrant fire organization did not have to evolve out of a suppression program. In places like Lake Wales Ridge it could make prescribed burning an extreme sport, as exhilarating as smokejumping. It made a good partner, ready to roll up its sleeves and do.

The key was land. The Conservancy was accepted not simply out of goodwill but because it was a landowner, with a landowner's rights, prerogatives, and responsibilities. It acquired lands by purchase just as other conservation lands in Florida did, and it managed lands for others (notably, DOD). It was not simply an environmental advocacy group like the Sierra Club, nor a research facility severed from operations like the University of Florida Botany Department or the Canadian Forest Service. It was a committed land manager who could not evade problems or wait upon research to suggest remedies or defer to legislatures and courts. But neither was it originally chartered to manage fire-type landscapes as Tall Timbers Research Station was. TNC burned to support its mission to nurture habitat and save species. It saw in fire a shared need and a collective threat. What remained unburned threatened not only its own holdings but its neighbors, what was burned badly by others could slop over onto its property. Whatever its founding expectations, it had to handle fire, and it had to learn how to do so.

What makes the Conservancy's Florida story so powerful is that the two models—Florida's and the Conservancy's—found ways to collude rather than collide. A fire strategy based on prescribed fire, a commitment to landowner rights, a determination for multiagency cooperation, a preference to be proactive—these were genetic properties of Florida fire agencies, yet they were also traits bred into TNC from its conception. Each emerged from that chrysalis stronger. TNC became a primary vehicle for translating the Florida model outside the pyric and political eccentricities of the state. Equally, its Florida phase gave the Nature Conservancy a second leg to stand on, even if its foot rested on a torch-equipped ATV. It toughened the institution's skills, kept it from being isolated into hobby burners, and helped carry its larger message about land, biodiversity, and burning to a wider world. After its Florida workouts, it punched well above its weight. It was the only private conservation organization anywhere that had in-house capabilities to manage fire.[5]

Even so, that alignment of tumblers could only fall into place and unlock the potential if there was someone skilled enough to work them. If, in those years, the Conservancy and the Florida model seemed to be everywhere, it was because Ron Myers was too.

It's seductive to anthropomorphize institutions. It simplifies analysis to treat as one entity what holds scores or thousands of refractory agents, and from a literary perspective, anthropomorphizing allows for a more direct narrative since institutions can serve as characters. Yet despite Supreme Court rulings that corporations are "persons," they aren't. What happens happens because people act or are acted upon; they ponder, fear, react, believe, stumble, leap, argue, fight; they choose. Often these acts get merged into a statistical composite that we call an institution. But sometimes the people involved are so few and they achieve so much that the institutional story and their own are virtually the same. That seems to be the case with TNC and fire. Jeff Hardesty, Paula Seamon, and especially Ron Myers took a Conservancy fire operation hardened in Florida and propagated it countrywide, and then overseas.

As the Nature Conservancy celebrated the 60th anniversary of its incorporation, the 25-year span during which it had stabilized its own fire needs and become a national firepower shrank in proportion to its longer history. The program that Seamon and Myers "grew" had grown up. The program had built capacity, it had adopted national standards, and it had internalized its lessons learned. The chapters could take care of themselves, a new directorate for the Conservancy decided. The Fire Learning Network replaced the national directorship as a means to link scattered holdings. The chapters could strengthen by bonding with their local cooperators. The national Coalition of Prescribed Fire Councils could institutionalize what had been held together by the strong nuclear force of personality. The Florida chapter assisted projects in Belize and the Caribbean. It appeared, too, that the new regime believed (hoped) that TNC could achieve its mission with less fire, that fire was not integral and universal, that Florida was an exception rather than a norm. What happened in Florida could stay in Florida.

It couldn't, and those expansive years in scrub pine and palmetto did not flare and expire like a flaming match head. They left behind a hard deposit that would endure, among which are lessons for all fire organizations about nimbleness, commitment, pragmatism, calculated risk taking, reconciling big ideas and big institutions with local knowledge and small communities. The experience bonded prescribed fire to ecological burning, not simply as a tool to reduce logging slash or as a sideline for suppression organizations. It showed that even a single person, without obscene wealth or inherited influence, could shape the national scene, the way a small switch can turn on a powerful dynamo.

When his tenure with TNC ended, Ron Myers stayed on the line. He had reached retirement age, yet wished to continue doing what he had done for a lifetime. He opened a consultancy. He kept his personal ties with Mexico. The U.S. Fish and Wildlife Service hired him to devise fire management priorities for the National Key Deer Refuge at Big Pine Key, about 25 miles from Key West and a hundred from Miami. One kind of temperament might see that as a virtual exile, fire management's version of Devil's Island. Another—Ron Myers's—could see Big Pine Key as the latest in a litany of protected sites, a tightly bounded place that needed fire and had little room for either error or dithering, a place where the Florida model met Latin America and the Caribbean. It was a place where a man with tenacity, conviction, and verve might continue to keep one foot in the black.

FIRE 101 AT STAR FLEET ACADEMY

Merritt Island National Wildlife Refuge

THE VISITOR COMPLEX at Kennedy Space Center is a theme-park paean to propaganda and the human presence beyond Earth. Two astronaut mannequins loom over the entrance. The rocket garden is a gleaming grove of those missiles (and only those) used to loft capsules stuffed with people. A robot explorer exhibit makes clear, through anthropomorphized mechanical "scouts," that those plucky spacecraft exist only as "trailblazers for human explorers." What is not said, what doesn't need to be said, is that the voyage beyond begins with a blast of engineered fire.

The visitor complex is where it is because it abuts NASA's major launch facilities. Those pads are at Cape Canaveral because being nearer the Earth's equatorial bulge grants a boost to rockets that more temperate zones can't deliver and because flights arc over the Atlantic, which can absorb a lot of failures. The Cape juts out so far because it is really two barrier islands, with the interior one, Merritt Island, separated from Canaveral by the Banana River and from the mainland by the Indian River, thus further isolating the installation.

These are the geophysical reasons. Had someone better versed in natural history pondered the site, he would have identified it as one of nature's great installations for pyrotechnics. There is more lightning in central Florida than anywhere else in North America; and more to the point, there is more lightning-kindled fire. Merritt Island is where lightning's ceaseless countdowns launch fire heavenward in boiling clouds of smoke. It is where, in Fred Adrian's memorable phrase, fire gets managed along a "wildland-galactic interface."[1]

When NASA decided President Kennedy's call to go to the Moon required an enhanced launch facility, it expanded, by purchase and condemnation, from an Air Force installation on Canaveral into Merritt Island. That gave it a lot of buffer zone, but one that needed an administrative presence. In 1963 NASA signed a memorandum of understanding with the Interior Department to have the U.S. Fish and Wildlife Service manage those surrounding lands. The landscape proved to be more than an inert swamp of gators and skeeters. It burned. The FWS soon realized it could not simply attend to eagles and migratory waterfowl. It had to cope with all the species of fire.

The fires were relentless. Although heaviest from May through September, the main storm season, lightning historically started fires every month save October. Meanwhile, fires from human causes occurred constantly. The refuge found itself fighting fires, and before long, lighting them in an effort to beat down some of the scrub that fueled the wildfires. NASA realized that its launch facility was very far from a clean room: fires burned around the base like feisty raccoons and smoke swirled like herons. Unless contained, they could threaten the infrastructure, halt launches, jeopardize instruments and optics that had to be quarantined for months (like the Hubble Space Telescope), and interfere with shuttle landings. Moreover, although seemingly constant, wildfires were unpredictable in occurrence and spread. Highly politicized, billion-dollar programs were hostage to the whims of giant electrostatic matches foraging the launch facility for kindling.

Yet, in those years, the agency was not the sophisticated fire operation it eventually became. It was itself overmatched. The agency knew a lot about ducks, egrets, and otters. What it knew about fire was local, personal, and institutionally unsystematic. Firefighting crews were "militias"—call-ups of local staff primarily hired as refuge managers, biologists, mechanics, or wildlife techs. They fought fires with surplus military vehicles outfitted with pumps, or with small tractors and plows, and with a lot of burning out. Prescribed fires were routine, even tedious affairs on small plots, a step above matches tossed into grass and subject to oversight by NASA's charged calendar. The agency stood apart from an evolving interagency fire community that pushed hard for common standards in equipment, safety gear, and training. In average years the staff got its fire chores done with no more fuss than it did the other varied tasks demanded of a refuge with too little money and too many calls on its time. Wildfires got knocked down, prescribed fires got lit. Yet year by year the amount that needed to be burned fell behind, a backlog that wildfire chewed on like turkey vultures on carrion. And then there were the exceptional years, the ones that accounted for most of the acres and crises.

On June 8, 1981, lightning set three fires, one of which shifted suddenly when prevailing southeast winds were overtaken by violent downdrafts from a thunderhead blowing from the west. The fire front's abrupt change surprised and overran a two-man tractor crew. Both men died from their burns. One was the son of a politically connected federal judge. Congressional hearings subsequently blasted over the Fish and Wildlife Service with the force of the fatal Ransom Road fire. The tragedy acted on the FWS much as the 1967 fire that consumed Apollo 1 did on NASA.[2]

The Merritt Island fire, following another fatality burn in 1979 at Okefenokee, brought money and a national fire program to the agency. Merritt Island NWR scaled up, its personnel and hardware becoming interchangeable with fire crews, equipment, standards, and practices elsewhere. In particular, it committed to prescribed burning as the best available strategy for containing wildfire. It faced the same suite of challenges as everywhere else in the national matrix of exurban sprawl and abandoned wildlands that went under that lame label, wildland-urban interface. But Merritt Island also had to deal with NASA, which was mightily irked when wildfires upset missions but was even less tolerant of fires set by nominal partners. In addition to the metastasizing checklists that everyone else had to consider, Merritt Island also had to seek approval from NASA, which was not keen on having long-planned missions to other worlds delayed by smoke on this one or having shuttle-weary astronauts remain in orbit while refuge fire crews burned wetlands and flatwoods to assist wood storks and the Atlantic salt marsh snake.

But NASA did not hold all the cards. There was a compelling case for doing the burning more or less on a schedule rather than leaving the task to lightning, arsonists, and off-road catalytic converters. No less, the FWS had the Endangered Species Act on its side; in fact, it administered the act. The Kennedy Space Center might sit next to the Astronaut Hall of Fame, but Merritt Island was a biotic chamber of threatened and endangered species (10 listed, 93 of state and federal concern), most of which depend on habitats shaped by fire. Especially noteworthy is the Florida scrub jay. Merritt Island holds one of the three viable populations in Florida, and the most vigorous one.

In 1993 when NASA wanted a new facility to support the International Space Station, it sought a site within prime scrub jay habitat. It got approval, but in return it had to support restoration of an equivalent patch of scrub jay habitat elsewhere on the island. The refuge staff built that landscape through industrial-strength slashing and burning. There was no alternative to fire: the fastidious jay not only demanded sand pine of a particular height but open land adjacent on which to forage. Lightning and scrub jays dealt the refuge two powerful trump cards.

This was not the classic wildland-urban interface (WUI), where fires start on one side of the fence and threaten the other and where one group wants fire (and accepts its smoke) and the other group does not. At Merritt Island each side has to accommodate the other; each provides windows for the other's operations. With coordination NASA can launch and land without smoke and flame as checks; what stalls its launches are the shuttle's flaws, not the refuge's fires. For its part, the refuge burns an average of 20 percent of its burnable land annually. It fires off the marshes every 18 to 36 months; the flatwoods, every 3 to 5 years; the scrub, every 5 to 10 years. It leaves unburned many sites adjacent to restricted facilities. Convective columns cohabit with rocket plumes.

Still, the geographic setting replicates—even if it exaggerates almost to parody—the national fire scene. On the cover of the refuge's 2003 fire plan there is an aerial photo that shows, in one snapshot, a launch pad to the north, an ordnance dump to the west, an explosive gas storage tank farm to the east, buildings and contaminated soil to the south; just beyond are restricted zones where deadly force is authorized. But then, as nature showed in 1981, it too can call upon deadly force against intruders. In the euphemism of the day, all this makes quite a challenge.

Among the theme-park schlock that infests the visitor complex, there is a show called *Star Trek Live* in which actors, under the auspices of Star Fleet Academy, dress in costumes, do skits, and explain the rationale for the human colonization of space. The show runs next door to an exhibit about what resources Mars offers colonizers and what life within a habitat module on the flanks of Olympus Mons would be like. Why a putative superpower would use a TV-show-turned-movie-franchise to promote a major expression of its ambitions and status, how one fantasy might be acting out another, is a question best stepped around, like meeting an irritated alligator at a pond.

It suggests, however, that the real wildland-galactic interface is more implacable than the garden-variety WUI because it combines commerce and politics with utopianism. The spectacle at Merritt Island slams together two incommensurable visions of the human future. One seeks to use controlled fire to leave the planet. It takes as axiomatic that the ultimate security and perfectibility of humanity lie in leaving Earth. The other proposes to use fire to make the home planet, the only one we will always reside on, more habitable. It accepts as a practical and moral charge to enhance what we have and to preserve its interstellar uniqueness. It assumes that our true destiny is not to understand how to live like hamsters on Mars but how to live like human beings on Earth.

UNDER THE DOME

Big Cypress National Preserve

IN SOUTH FLORIDA the landscape further flattens (if that can be imagined), like an intricately carved candle placed in the sun and melting into a widening pool of wax. There are fewer ridges and hills, and more prairies, marshes, sloughs, and glades. Topographic relief is measured in feet, and then in inches. Lake Okeechobee is 13 feet above sea level; the grade over the Everglades is two to three inches a mile. What happens to land happens also to rivers, which spread into thin sheets. The texture of the scene resides in the contrast, often striking, between flat grasslands and mounded clusters of trees, variously grouped as pine islands, hardwood hammocks, or cypress domes. Geology morphs into biology. The landscape becomes a bioscape.[1]

Why two biomes so different should coexist with such tenacity remains an ecological mystery. The dynamics of each by itself are understood; what puzzles is the nature of their interaction because, while the grassy swathes burn routinely, the mounds burn rarely. Without fire the prairies, glades, and seasonal wetlands redolent with rough are overrun by woody species and may in turn become hammocks or domes. If burned, however, the hammocks or domes dissolve into muck and may become glades. What is inexplicable is why those fires—and the region is blasted by burning—stop at the green fringe of hammock and dome. The border between them is a line drawn in the limestone.

The customary explanation is that they stop because they can't continue. There is something in the biophysics of their setting that increases fuel moisture, that blunts wind, that favors more incombustible fuels. Even under extreme droughts, flame struggles to creep over the surface or seep down into the peat. The pine islands sit atop slightly elevated limestone that can break the flaming front. The hardwood hammocks grow on a rising pile of peat. Those organic soils are just too moist and compact for the ephemeral flames from prairie or marsh to kindle. The cypress domes organize around holes in the limestone that fill like ponds; the trees ring them, spreading outward such that the oldest and tallest are closest to the hole, creating what appears from afar as an organic tumulus. Because they can draft from the pond, their fuel moisture remains high and they can resist the flash of an approaching headfire. The bioscape burns, yet keeps its structure. The fires only sweep away the annual debris, like a spring freshet that flushes its channel without relocating it or a wind that blows fallen leaves without uprooting the forest.

This peculiar biogeography of wet savannas and clustered trees is documented in the earliest records, so the arrangement must be natural, or have a powerful basis in natural conditions, and it appears that the patchiness works to maintain itself, the one to encourage fire and the other to defuse it. The assumption is that the ultimate explanation must lie in the physical matrix of rock and climate, of which the living world is only an epiphenomenon. But analogous fire landscapes can be found in tropical and subtropical savannas in South America, Africa, Asia, and Australia in which there is no comparable hydrology, yet the fires rage until they strike the green wall of woods and suddenly expire. Again, no adequate explanation exists.

It may be that a better answer lies not in fuel, moisture, and wind—the biophysical basics of fire behavior models—but in bioecology. A fire in a furnace is a mechanical process that consumes fuels. A fire in a landscape is a biochemical catalyst that interacts with its surroundings; while not itself living, fire is a creation of the living world. A firestorm depends on life in ways a thunderstorm or a hurricane does not. What may be missing in the texture of the south Florida bioscape is to think more imaginatively of its fires as a process that has coevolved with the stuff it burns, or more properly with all the flora and fauna that also shape the stuff it burns. Fire interacts with its biological surroundings as much as with terrain and wind. And that biological leveraging is what may be lacking in efforts to restore fire regimes by flame and fuel alone.

In Africa elephants trash trees, consigning limbs and small trunks to the flames, and so help maintain the balance of woodland and savanna. Similarly, porcupines, by injuring young trees and forcing them to exude sap, give fire a needed leverage to limit the thickening of woods. By itself fire cannot break down or drive back the woody clumps; by interacting with other creatures, it can. It may be something analogous is happening to shape the south Florida scene. Black bears feed on palmetto hearts, and their decline may help explain why fire alone cannot pound palmetto down to its historic scale. Alligators clean out the center ponds of cypress domes, which limits the buildup of peat. The pine straw, the peat, the muck, none fall from the sky like rain or ooze upward through the limestone like a rising water table according to gravity and pressure. They are the outcomes of living organisms, or of physical processes absorbed into and reworked by biological agents.

The domes set up positive biotic feedbacks that push out and stiffen their edge. The mounds rise, thicken, and perhaps slow their ecological physiology. Occasionally, when drought is severe or the wetland drained, the edge effect crumbles and the domes and hammocks burn, a kind of pyric plague that sweeps away and freshens the site the way that rare but essential crown fires blast through old-growth Douglas fir stands or mountain ash eucalypts. Instead of a flashy jolt, a slow burn pries loose the pieces and allows them to reorganize. At such times edge effect is etched across history rather than geography.

Because so much fire science involves fire behavior, however, even fire ecologists tend to restate their biotic landscapes into the conceptual language of fire physics. They assume the biological effects follow from the physical. The south Florida bioscape suggests that just the reverse might be more accurate. The physics of fire behavior may be a secondary outcome of fire's biology.

That is the natural (or quasi-natural) landscape. Curiously, the contemporary built landscape has taken on a similar texture in which small enclaves dapple a broader plain. In this case the hammocks and domes are nature preserves poking up amid savannas of suburbanization. Again, the two systems argue for incommensurable fire regimes in which flames stop, or must be made to stop, at their borders.

In the 19th century a train of legislation oversaw booms in timber, cattle, and citrus, along with a good dose of speculative land fraud. The key was to dry wetlands and fill them with newcomers. From the federal Swamplands Act of 1850 to Florida's own Internal Improvements Fund (1879), the intention was to convert unproductive wildland to revenue-yielding rural land, or at least "improved" land that could be flogged off to unsuspecting investors. In south Florida scheme after scheme focused on draining, and for a long time the swamps won. They were too vast, the land too mucky, the applied technology and capital too sparse. Then the plugging and the rechanneling took. The regional hydrology got replumbed. Some landscapes were changed usefully; many just sprang leaks elsewhere or began wasting away as though from ecological scurvy.

The modern era began with the need to pick through the wreckage. Unpatented public lands in northern Florida were made national forests. Sanctuaries were established for migratory birds. And when lands fell delinquent in taxes during the Depression, the various levels of government, but primarily the Feds, again stepped in not only to remove the land from further abuse but to rehabilitate it. In fact, one argument for acquisitions was to have sites the CCC could work on. The Florida National Forest added Ocala; Florida got its first state parks and forests. Other than posted signs, however, the old ways often continued, with the land used by hunters, leased to ranchers, and cut by timber companies. The big edge effect came with the postwar boom.

What had previously been drained of water now flooded with people. Interstate highways replaced cross state canals. The exurban frontier remade the landscape more ruthlessly than clear-cutting. Developers stamped block after block of suburbs, shopping malls, office parks, and golf courses. This time the boundaries were not permeable: the border between town and country was as sharp as a hammock's edge. In response, a counterwave of legislation, an inverse of that which had enabled the old frontier, arose to halt the encroaching houses, theme parks, and consumer sprawl. While schemes that sought to "manage growth" outright failed, a bouquet of smaller acts allowed governments at all levels to buy packets of land for nature protection. The genius of the program was to finance the bond issues behind the purchases by taxing real estate transactions. It was as though, in the 19th century, logging companies had underwritten forest reserves.

An astonishing rush of acquisitions has resulted. By 2006 the state had purchased over six million acres for forests, parks, and wildlife conservation; it had an acquisition budget of $300 million, "far exceeding that of any other state or even that of the Federal government for use in all fifty states." The euphoria spread into counties and even cities; 29 of Florida's 67 counties have analogous programs, and have raised over $2 billion to purchase 375,000 acres. Combined with federal holdings, Florida has 30 percent of its landed estate in public land. What the federal government had done in the late 19th century, the state government was doing in the late 20th, this time seeking to shield from asphalt instead of axe.[2]

The incentives, as should be expected, were many; after all destroying its natural scene would not encourage more immigrants or tourists, the basis for Florida's economy. But biological considerations were powerful. Florida did not offer Alpine scenery or a Hudson River pastorale, it could not match Yellowstone's geysers or Grand Canyon's gorge: its geology, like its rivers, was subterranean. What it offered were green strata of ecology, diminutive mountains of hammocks, flocks of flamingos rather than bison, and a tallgrass prairie of saw grass. The first wildlife refuge had been proclaimed at Florida's Pelican Island in 1903 to protect plumage-rich species from hunting. When the Everglades was declared a national park in 1947, it was the first in the system to enjoy protection for its scenic biology rather than its rocks. What drove the postwar era, however, was an interest in biodiversity and the immense leverage offered by the Endangered Species Act passed in 1973.

The red-cockaded woodpecker and scrub jay were the most famous firebirds, but neither could match the Florida panther for charisma. Crowded by the postwar frontier, the far-ranging cat had retreated to the more loose-limbed landscapes of south Florida. There its presence boosted efforts to halt, and maybe even reverse, the deranged water regimes that had evolved. Between the panther and the need to stabilize the regional hydrology, protected lands spread out from Everglades, like a reverse sheet flow. Florida's state parks got Fakahatchee Strand Preserve; the Florida Forest Service, Picayune Strand State Forest. The FWS got the Ten Thousand Islands and the Florida Panther National Wildlife Refuges. North and east the drive for protection slammed against the Seminole and Miccosukee Indian reservations and the amoeboid concrete of metropolitan Miami. But the whopper, the keystone acquisition, was Big Cypress in 1974. A shallow basin, it became itself the Big Cypress Dome of Florida environmental history.[3]

That was the political geography, an uneven dialectic between development and preservation. Land logged, plowed, grazed, even drained, all could, at costs, be rebuilt or abandoned and gradually shuffle into something that had a semblance of its natural origins. Land scalped into office parks and platted suburbs could not. The starkness of the contrast gave urgency to the acquisition program. Once cut over, longleaf might be replanted to slash pine; but once paved, the land was lost. Modern Florida became a dappled landscape, not with wetlands and ridges, but with urban patches and parklands. In south Florida the contrast resembled a macabre parody of the natural order, rumpled with suburban savannas and nature-preserve hammocks.

Shielding places from urban conversion, however, did not save the biota. It could be lost from neglect as surely as from abuse. The ability to maintain was as critical as the capacity to purchase. In Florida that meant the capability to manage fire. Don't buy if you can't burn.

Wildlife and water were the legal reasons for the reserve. But replumbing the water regime was an immensely expensive, long-term project, and organisms had been accommodating the changes over nearly a century. The immediate reality was fire. Left unburned, the habitat became unusable by panthers in as little as four to five years.

Fortunately, historically, there had been plenty of fire. What lightning didn't burn, people did, and vice versa. The biota could take a lot of fire of all kinds and in all seasons. It seemed, in fact, remarkably resilient to burning. What it could not survive was fire's absence. Fire exclusion would kill off the panther, and much else of the living landscape, more surely than poachers. Today, Big Cypress National Preserve burns an average of 50,000 to 60,000 acres annually from prescribed fire, which is more than the rest of the National Park Service collectively. A national agency with 334 sites and 83.6 million acres has half its prescribed fire acreage burned by one unit. If Everglades, which adjoins Big Cypress, is included, that figure bumps up to 80 percent. Industrial strength burning on this order reduces the rest of the national park system to boutique burners—mom-and-pop shops competing with Walmart.

So fires matter along the borders, where the park abuts private lands. In the early years, the fire dynamics of the fringe resembled that between hammock and grassland. Free-ranging fires raced against those boundaries and threatened the preserve; aggressive firefighting sought to hold the line, to make the fire border as precise as its legal one. But in recent decades, as public land was being purchased rather than sold, the pyrodynamics have reversed. The fires flourish within the legal hammocks and threaten the housing tracts outside the border; aggressive fire management seeks to hold them within, or to substitute tamed for wild fire. By adding units, all levels of government in Florida have multiplied the borders, which is to say, they have increased the problem of keeping fire in where it's needed and shutting it out where it isn't. Occasionally the system crashes, and the outside burns catastrophically. Sometimes the disaster renews the politics. Sometimes it adds to the wreckage.

What nationally is (fatuously) known as the wildland-urban interface has a sharpened sense in Florida because fires are not only so ferocious but so frequent. But what makes the Florida scene curious is that, unlike most places, where the border must be impermeable to fire, some places here are porous—and that holds for definitions of fire. Elsewhere, fires are either prescribed or wild, wanted or unwanted, natural or anthropogenic. The definitional border is as sharp as the edge of a pulaski. But in south Florida, in particular, the boundaries between space and time, political geography and fire regimes, blur. They have certainly blurred at Big Cypress National Preserve. Fires wash over borders and cross categories much as waters fill up and empty the shallow basin. The borders are too low and too unstable to hold.

Big Cypress collects, holds, exaggerates, and exemplifies. It is a chimera of a park; part natural preserve, part recreational site, part cultural reservoir of traditional sites and practices. Like the pine conversion projects at Ichauway Plantation, it has involved a slow transformation of the land, over many decades, that would manage to keep its forms while replacing its functions, and where old functions remained, it would repurpose them.

Compared to a pristine site, the place was a clutter of the good, the bad, and the ugly. It was a landfill of junk practices, full of hunting cabins, free-range cattle and open-range off-road vehicles, oil drilling pads, landing strips, the Dade-Collier Jetport, radio repeaters, hydro stations, the Florida National Scenic Trail, I-75's Alligator Alley, the Tamiami Trail (US 41), archaeological ruins, a score of American Indian villages, and places of significance to neighboring tribes. Its old-growth cypress had been cut out. Its cypress domes were sad diminutions of once-towering giants. Lawlessness had ruled in the past, and it returned in the form of drug trafficking trails and drunken weekends in the woods. Yet jumbled into this hodgepodge were 120 plant and 33 animal species threatened, endangered, or of special concern—Cape Sable seaside sparrows, red-cockaded woodpeckers, indigo snakes, Florida panthers, eagle nests, patches of relic old-growth cypress, and a sloppy biotic shoreline where a subtropical biota ebbed and flowed against a temperate one. Its fire regime shared the shambles. There were fires set by lightning, fires set by cattlemen, hunters, and traditional users, and fires kindled by accident or arson. All flowed in and out of the shallow Big Cypress hollow.

That's why Big Cypress was designated a national preserve, not a national park. Gradually, it began to convert from that invented status to more traditional forms. It bought out the inholdings. It disabled all but two of the drills. It protected the nesting sites. It fenced the open range from off-road vehicles, the mechanical equivalent of cracker cows, putting them into regular trails. It extinguished the grazing leases. It shed many squatter camps and shrank the remaining hunting camps to a few acres. Many of the traditional activities continued, though less riotously, but many morphed into surrogates better suited to a nature park—from hunting to bird-watching, from swamp buggy off-roading to hiking. The preserve imposed a rude administrative presence that had never before existed on what had been largely a private fiefdom. It shielded the threatened and endangered species. And it took the torch from the hands of freewheeling folk and gave it to agency burners. That in itself was not unique: the identical transition had happened all over Florida. The preserve substituted official fires for folk fires the way giant cypress replaced longleaf.

What moved the Big Cypress experience beyond the realm of what was ordinary, if huge, was the multiple ways the exchange had occurred. Hunters replaced ranchers, and then tourists replaced hunters. Fires lit to stimulate browse tweaked into fires lit to green up habitat. Where hunters had burned for deer and turkey, fire crews burned for panther and Cape Sable seaside sparrow. Where loggers had fired the slash, fire techs burned around nesting trees to protect them. The statistical record of fire that elsewhere divided between wild or tame, lightning or human, here fell into three categories: natural, prescribed, and anthropogenic. The process of conversion would be a long one; the categories would mingle and split only after many years; and likely they would never self-segregate completely, a condition new federal fire policy encouraged as it accepted fire as fire and judged it by how it affected the land. The fires at Big Cypress merged and overlapped and moved in and out of conceptual boxes that elsewhere seemed as impermeable as the legal borders that segregated park lands from private.

The preserve's headquarters at Ochopee are a reconditioned hotel, complete with swimming pool, in which former rooms have become offices. Its Fire Operations Center is run out of a former townsite. That, in a nutshell, is what is happening to the management of the preserve's natural estate as well. Its fires are reoccupying and redirecting those that have shaped Big Cypress over the previous century. Because of aggressive land acquisition, the place is buffered against the worst expressions of urban clear-cutting, and because of its late and complex history, it is buffered as well against the sudden inversion of fire regimes that has upended so much of contemporary Florida. Its fires continue to forage, not unlike the panthers.

Only minuscule rises define the borders of this shallow basin: it is porous to water, animals, and fire. Yet even small differences, over time, can reinforce themselves in ways that make a patch of peat or a sinkhole into a hammock or a cypress dome. The processes underway will have the chance to continue. As decades pass, the distinctions can heighten, the edges sharpen, and Big Cypress will thicken into the giant cypress dome of south Florida nature protection. It will look and operate differently from the lands outside. Likely, those lands outside will show less fire or more wildfire. Under the dome the fires will flourish.

THE EVERBURNS

Everglades National Park

THE FLORIDA STORY flows south.

It ends when the land spreads into a wide delta of limestone and sedge rimmed by mangroves and sea. At the Everglades water, fire, and people converge to make, in Marjory Stoneman Douglas's famous dictum, "one of the unique regions of the earth," a place like "nothing anywhere else." Here, subterranean aquifer and surface overflows merge into a single stream; and here the fires set by lightning and those by people, the regimes hidden by evolutionary time and those visibly spilling out from the flow of settlement, converge. The pieces that make up this scene are common to Florida generally. It is how they come together and heighten that make the Everglades singular.[1]

Everything shrinks to a minimum, as though the Earth's landscapes were infinitesimals approaching a limit. Relief is measured in inches; peat domes and sinkholes replace ridges and swales. The dense biotic mosaic of the subtropics thins into patches of pine rockland and hardwood hammock on a sea of saw grass, like atolls in the Pacific. Rivers become sheet flows moving only hundreds of feet in a day. Ecological dynamics dissolve into a dialectic of water and fire, rising and falling, coming and going, flickering like Schrödinger's cat from the fabled river of grass into a river of fire. The Everglades could as easily have been named the Everburns.

Or it once could. Like Florida generally the Everglades is a broken landscape. A century of hydrologic engineering and ragtag plundering have shattered the rhythms of flooding that annually passed through the

Glades. The waters that flooded hummocky landscapes and once spilled south from Lake Okeechobee to pass over the land like thick ooze have been diverted into canals for agricultural fields and cities; have been halted by ditches, roads, and developments; have been drained away or dammed or released at the wrong times or come laden with toxins. The regularity of fires broke down in the same way. The fireshed fragmented. Fires were too few, or too intense, or out of sync. Water and fire, each emulated the other. Each needed the other.

The Glades's depth relied on its breadth. The rhythmic flows of water and fire depended on a wide watershed and a broad biota that could absorb and buffer. Like a graded stream that adjusts its profile to accommodate water to sediment, the Everglades's two flows adjusted their profiles to the amount of work required, the one reconciling water to sediment, and the other flame to fuel. But settlement ripped away those borders, flaked off chunks of the biota, and slashed its size in ways that left it starved for water, stripped of critical species, and exposed to fires that could burn down instead of out.

As the water fell, the fires burned deeper and spread wider. The canals and roads that water had dissolved previously now disintegrated through muck burns. In 1920 hammocks that had for most of a century provided sanctuary for Seminoles burned away. Ernest Coe, who began campaigning for park status in 1928, argued that "fires have always swept the Everglades and they are going to continue to do it." But these fires had become pyric mutants: they didn't burn in the old ways. In 1935 soil and water conservation authorities successfully urged the legislature to establish a fire control district. It had mixed results, swatting out small burns, though proving ineffective against frost and drought (or its artificial surrogate, drainage). When Daniel Beard conducted his wildlife survey in 1938, he estimated that over the past year fire had burned half the piney woods, 80 percent of the Everglades prairie, 30 percent of the coastal prairie, and 5 percent of the Ten Thousand Islands Coast and "about the same amount of the Cypress." In 1941 a fire "of incendiary origin" blasted over 250,000 acres and left crews scrambling to get out of its way.[2]

In 1945 the full magnitude of the tinkering became unblinkingly clear. Drought settled over the Glades, as it had for millennia, but this time the wetlands had lost their capacity to shrug off the drying, to sink into deeper holes like its gators and wait for the rains and spillage to return. Surface fires burned to the new water table, where they smoldered wretchedly through peat for days, for weeks, spewing an apocalyptic pall of smoke that gagged the region. The river of water shriveled to a trickle; the river of fire flooded, its flames scouring its historic channel down to bedrock. In Douglas's outraged words, "The Everglades were dying." The Indians, with "the stoic faces of fatalism," saw the "end of their world." The white man, at last, realized the consequences of his fecklessness. For the first time the "problem of the Everglades was seen whole."[3]

The Everglades is a landscape of ideas as much as of egrets and thunderstorms. Behind the hand that worked guns, axes, and dredges lay a mind that imagined what the place meant, what it might be good for. In this sense, again, the concepts, beliefs, and tropes of imagination that guided fire practices in Florida also flowed into the Everglades. Here philosophies of fire management converged with a special clarity made possible by the apparently simplified landscape and its intensified scrutiny by outsiders. How the Everglades coped with fire became a cameo of Florida fire, and of American fire generally.

The great burn of 1945 gave urgency to the movement to legally protect the Everglades from further abuse. Two years later the Everglades became a national park, the first established primarily to protect biological values. The wretched peat fires had acted on south Florida as cutover slash had in the Lake States 50 years before, kindling national outrage and leading to schemes to protect "from fire and axe," or in the Everglades, from fire and dredge. From the onset, fire protection was fundamental to the park's purpose: the issue was literally existential. If rangers could not prevent the return of fires like those of 1945, there would be no park.

But ideas changed park fire regimes as much as they affected its hydrologic regimes. Those ideas came from the outside and had to express themselves amid the unique setting that was the Everglades, where, no less than other settlement schemes, they sparked unintended consequences. The narrative of Everglades fire is thus a chronicle of national notions meeting local circumstances. Those notions involved fire control, fire management, and fire restoration; the circumstances were a large patch of pine upland (Long Pine Key), expansive sloughs and wetlands, and a broad fringe of marshes and mangroves. Meanwhile, the park's borders changed as it absorbed inholdings, acquired additional lands, and confronted intense urbanization to the east.

There was no avoiding fire: it was endemic, frequent, and indomitable. It would come even if people vacated the place. Saw grass burned over standing water, hardwood hammocks burned in drought, mangroves burned after frosts, and slash pine rough could burn anytime. A subtropical place in which something was always blooming meant that something was always burning. Equally, there was no avoiding scrutiny. The park resided next door to that urban conurbation known as Miasma (aka Miami), while its iconic status made it visible across the country and vital to national conservation organizations. Inevitable fire, implacable attention, a fire park in an agency undergoing reforms in its fire policy—the Everglades became an anomalous leader in the national story.

Initially this meant fire control. Ending the burning was a reason for the park's reservation in the first place. As its first superintendent, Daniel Beard put it, "Under present circumstances" fires of "all kinds" must be "prevented, or extinguished if they start." In fact, aggressive firefighting, perhaps leading to fire exclusion, had long been a policy of the National Park Service, stiffened by the Forest Service's adoption of the 10 a.m. policy in the spring of 1935. Both the policy's target time for control and its purpose had little meaning in south Florida. In the Northern Rockies, where the policy was birthed, 10 a.m. marked the breakup of the evening inversion, which led to quickened burning; this had little pertinence in a tropical landscape as flat as a bureaucrat's desktop. Nor could fire be excluded. Lightning and people had burned the place as often as it could carry fire. Daniel Beard, whose 1938 report for the Interior Department provided the scientific basis behind park legislation, and who became the park's first superintendent, began his observations on fire by repeating the old saw that "Florida burns off twice a year." It was, he concluded, "hardly an exaggeration."[4]

Still, the NPS commenced with a control policy that, as the park's chief scientist noted, consumed much of the staff's energy but did little to alter the fundamentals. Meaningful fire control was impossible without water control, and until the park was reflooded, its staff could only swat fires as they might mosquitoes. Newcomer rangers were appalled at local fire brigades who simply burned out from the nearest road; it was better, they thought, to attack the fires directly. But bulldozers, plows, pumpers, tracked Bombardiers and Thiokols, even glade buggies, too often tore up the landscape (and got themselves torn up or mired in muck). Fires burned across deep-pitted pinnacle limestone, through decadent saw grass overtopping knee-high standing water, amid coarse subtropical rough that could carry fire any month. Tactics used elsewhere proved as helpless as pulaskis; and ideas floundered as much as tactics. Not least, change happened quickly. The effects of fire control in Oregon or Minnesota might take three or four decades to become apparent. In south Florida the outcome was visible in three or four years. Even casual staff, cycling through the park for a careerist tour of duty, could see the consequences.

In 1951 park biologist William Robertson began an investigation into what specifically fire did in the Everglades. Completed in 1953, "A Survey of the Effects of Fire in Everglades National Park" became an instant classic, one of half a dozen seminal fire studies in the century. While he concentrated on the slash pine uplands, whose dynamics most resembled Florida fire ecology generally, every part of the Everglades system had its fire regime, and a nationally based Park Service policy was wildly out of sync with all of them. Removing fire was damaging the biota the park was intended to preserve, while actual fire fighting with tracked vehicles and plows left lasting scars. A 1956 Fire Control Plan sought to contain fire but also to limit the damages mechanized firefighting did by restricting their use and to replace ad hoc firelines with a system of permanent roads.[5]

Behind the appreciation for the special difficulties of fire control lay the prospect that prescribed fire might be necessary as a surrogate for wildfire. In 1958 Everglades received authorization to conduct controlled burning in its pinelands, the first and sole exception within the national fire program. In 1962 wildfires swept much of the park; the next year, 1963, the Leopold Report catalyzed a shift in NPS fire policy, proposing that parks generally should be "vignettes of primitive America" and that any restoration would have to include fire. The report culminated in new guidelines for natural area management published in 1968 that sought to reintroduce fire where it had been lost and where ecosystems had suffered in consequence. Everglades was again at the forefront. In 1969 it expanded its prescribed fire program into saw grass prairies. Soon afterwards it sponsored research by University of Miami professor R. H. Hofstetter to update Robertson's classic.

Twenty years after Robertson's report, Hofstetter issued his own, complete with a long litany of recommendations. These were largely incorporated into the revised Everglades Fire Management Plan of 1973, further amended in 1974. As "fire management" replaced "fire control" in the title, so the intention of the plan was to "restore" fire to something approximating its "natural" or presettlement state. To this end, it parsed the Everglades into three biomes, each with its own prescription for burning: a coastal belt of marsh, prairie, and mangrove; the sloughs of saw grass and wet prairie; and the pine rocklands. Some fires would be left to roam, whether kindled from lightning or people. Many would be set deliberately. And some fires would be suppressed, particularly those when drought pushed the water table too low or when fires threatened inhabited borders; but even here "control" could mean a range of responses from direct attack to loose herding or backing off to barriers from which to burn out. Other parks were adopting similar strategies, but none could approach the Everglades for actually getting fire on the ground or through the sedges. The plan was approved the same year massive fires again descended.

While the program seemed radical on the national arena—and received prominent attention—it actually nudged the park into practices that had long characterized south Florida. A few problems were pragmatic, like keeping smoke out of Miami and fire out of deep-peat hammocks. But many were metaphysical. What was the "natural" fire regime? Should it include the anthropogenic fires that had coexisted with the emergence of the Glades over the past 5,000 years? And what did "natural" mean when the hydrology had been utterly replumbed, when the rhythm of fires had become atonal, when flammable exotics like melaleuca and Australian pine threatened to overrun burned areas, when there was always a Cape Sable seaside sparrow, a Florida panther, or an indigo snake that was endangered depending on how a place burned? It was simple to imagine untethered lightning fires reinstating a regime in the High Sierras or the remote Gila Wilderness. It verged on scholasticism to argue the merits in the vast marshy terminus of a peninsula that had had its indigenous character erased, its flanks crowded with avid invaders and a giant megalopolis, and its fundamental processes rewired, polluted, and deranged. The one axiom was fires would happen. They would come whether or not people could reconcile the thesis of the past with the antithesis of the present into some kind of synthesis for the future. The place would burn.

So the debate continued, one of many environmental themes at Everglades that captured national attention but had a very particular meaning in this very peculiar place. After all, how many universals could play out in a unique setting? The Everglades had an unblemished record of frustrating grand schemes with perversely unexpected consequences. Still, the Sierra parks and the Everglades made the odd couple of NPS fire management, and they attracted some of the premier fire officers in the service, with a few like Larry Bancroft working in both. For the NPS it was an era of exhilarating experimentation that only slowed, not ceased, in 1979 with efforts to standardize and nationalize practices. The park fire program flourished.

Then came the Yellowstone conflagrations of 1988. Even as the episode forced "prescribed natural fire" programs to cold start, which meant parks and forests had to redo and resubmit their fire plans, Yellowstone's potlatch brought huge sums of money and attention to fire management in the Park Service. What had been for most parks a tangential program, not easily put into visitor services, muscled into prominence. The outcome for Everglades was a sophisticated rewrite of its fire plan that modernized terminology, accommodated the boost in resources, and absorbed another management unit—a block of newly acquired land to the northeast (with actual purchase scheduled for 1995) that would regulate the water flow through Taylor Slough but enormously complicate the fire program because it included agricultural (and Hispanic) squatters, along with fireweeds like melaleuca and because it would press against the ravenous sprawl of the Gold Coast. The plan sustained the ambition to restore "natural fire regimes," although it craftily sought to finesse what this meant in practice. The essence was it sought to get more of the right kind of fire on the ground.

Until Big Cypress National Preserve arrived to make a tag team, no one in the NPS did that with more routine success. Even so, no one was content with how much got burned, and all worried about the increasing constraints imposed by outside interests and ideas. It was as though the invasive Burmese python had the park fire plan in its coils. With every operational hiccup or review-inspired gasp for breath, with every call by researchers for another study or reconsideration, or with every new advocacy group for a cherished species, the coil tightened. The process only went one way. Fire officers could hold, not expand, operations. Still, as fire management commanded more national attention through the 1990s, the program got additional money and personnel. The debates raged incessantly about how, when, and with what restraints to do the burning; but the burning went on. Compared to Everglades and Big Cypress, the rest of the National Park Service looked like a gaggle of hobby farms.

The arrival of the National Fire Plan and the Comprehensive Everglades Restoration Plan both bolstered and hobbled the Everglades fire program. The park enlarged and enriched its fire staff, but that effort dimmed in comparison with the $8.6 billion, 30-year project to reinstate more of the historic hydrological regime. Once again, water and wildlife dominated the scene, reducing fire to an epiphenomenon, a looming disaster that apparently could only be set right by a restored hydrology, or a threat to a swelling population of endangered species, each of which had sharp-eyed (and sharp-tongued) citizen advocacy groups peering across the park boundary. Fire had no such constituency; it had no Endangered Process Act to give it legal standing; no park had been specifically chartered to showcase fire. The park staff included one fire ecologist but, as a fire crewman shrugged, seemingly "hundreds" of hydrologists. Although a revised fire plan, accompanied by a formal environmental assessment, was headed for approval in 2011, in many ways fire's status remained where Marjorie Stoneman Douglas had left it 60 years previously. It was seen as a potential predator on the park, worsened and made prominent by the ruinous heritage of mismanaged water that threatened to drain the river out of the River of Grass. At best, to many laymen burning seemed ameliorative; it did not seem essential.

The larger fire community saw it differently. However hobbled matters appeared within the park, when viewed nationally the fire program at Everglades had been, from its origin, a powerhouse that helped move Florida fire from exceptionalism to exemplar. Everglades had sponsored the first comprehensive research on fire ecology, had conducted the first authorized controlled burns, had created the first prescribed fire program, had executed some of the earliest prescribed natural fires, had constantly challenged the National Park Service, and through it the national fire community, to rethink its premises. The Everglades idea pressed against its surroundings, and against all reasonable odds it had helped convert adjacent land into national preserves, wildlife refuges, state forests, and water conservation zones and had helped share their fire programs. No one doubted the significance of fire to Everglades. What was uncertain was the value of the Everglades experience elsewhere.

Within the national park system Everglades remained an anomaly. But then that nominal system was itself anomalous. Unlike other countries, the United States did not have a national park act that established new parks under a common policy. Instead, Congress or presidential proclamation established each park or monument with a separate act and purpose; what united the parts was a common agency to administer them. Even so, Everglades sat awkwardly in the national economy of fire. Ideas imported from elsewhere failed to take or like invasive plants they inflicted damage; and practices invented at Everglades struggled to survive when exported. Its fire program, that is, was much like the park.

There were two ways to characterize this outcome. In one, the place was sui generis, a fire autarky. Only in the most abstract and academic senses did it relate to fire elsewhere in the country. In the other, Everglades represented a valid if unquestionably unique synthesis of national ideas and local circumstances. It was as though the entire country had flowed southward into Florida and then stopped and congealed at the end of the peninsula. In this configuration Everglades was synecdoche for fire in Florida, and maybe more.

Long narratives can distill into people as well as places. At Everglades it gathers with uncanny appropriateness into the career of Rick Anderson.

His forbearers were among the wave of Scots-Irish that washed ashore at Oglethorpe's Georgia colony where they hoped to work off their debts and return to Ulster. Neither happened. In 1732 the Andersons and the McClellands first married. Then the clans began a slow, long trek south through the advancing marchlands of the Creek and Seminole wars. By the 1830s they were around Gainesville. They continued to probe south down the Lake Wales Ridge. The Andersons settled mostly onto farms, the citrus industry helping to root them. The McClellands yielded to frontier wanderlust and their herding heritage. Southwest Florida was still raw in the 1940s when his grandfather logged and hunted in the Big Cypress Swamp.

In 1957 Fort Myers offered the closest hospital, so Rick was born there. Mostly he grew up amid citrus and cattle near Dade City farther north, spending much of his time around Green Swamp, and always he heard the old family stories about life on the land. There were fires everywhere: they were just there, like the clouds and the summer thunderstorms and the piney woods. He grew up around burners and with stories of burning humming in the background like the summer's mosquitoes.

But the boom was on. Old Florida was vanishing: the longleaf were cut out and the big cypress gone from Big Cypress, the tick-ridden and free-roaming cattle were dipped and fenced, and the free-ranging fires increasingly prescribed by law and plow lines. In 1976 he joined the Florida Forest Service, right out of high school, eager, as he puts it, to find a job that would pay him to work with fire. The monster Turner #10 fire at Big Cypress found him among the futile suppression crews and brought him into contact with the National Park Service. He quickly realized that the NPS paid far more than FFS and transferred. Then he learned that being a part of a national agency meant he could go elsewhere in the country.

He went to Yellowstone in 1985 and was there during the great firefest of 1988. The experience left him full of questions: he'd started his career with a megafire, but the Yellowstone plateau and the Absarokas were vastly different than the Big Cypress and Lake Wales Ridge. Park researchers tired of answering his endless queries. Go to college, they told him. He was too old, he replied; he'd be in his mid-30s when he graduated. He would be in his mid-30s anyway, they answered. So in 1989 he enrolled at the University of Montana, which had a sterling fire curriculum and roosted among one of the country's great fire cultures. For two summers he went to Belize for research, and used that experience to segue into a master's degree in ecosystem management. In 1994 he returned to a National Park Service still flush with its post-Yellowstone funding.

He transferred to Olympic National Park, a cold swamp that, unlike Big Cypress, still had most of its giant trees but its only fires were in campfire rings. Within a year he moved south to Saguaro National Park, outside Tucson. The park ranged farther over the Rincon Mountains than the saguaros on its midslopes. It looked oddly familiar: this was a drier version of what he had grown up with, the western yellow pine replacing the southern, the muhly and fescue substituting for wire grass, summer lightning kindling fires like a Fourth of July picnic. What the Rincons had that central Florida did not was real relief: it had slope and a crenulated texture that created a biotic mosaic that in Florida had depended on differences in soils and water tables. He quickly built up a prescribed fire program that pioneered landscape burning, that used terrain to direct and contain long-burning fires, igniting at the summits and letting the flames work their way down. It was a top-down system built from the bottom up. Where, nationally, prescribed fire modules were still in the garage or spun wheels in bureaucratic mud, he found ways to get them on the road and give them traction. He worked around concerns with the Mexican spotted owl and a wary Forest Service—nothing here was as complex as it was in Florida.

Then came a prescribed fire brouhaha as great as Yellowstone's natural fire season. On May 10, 2000, crews at Grand Canyon kindled a prescribed fire that blew out of control and forced an evacuation of the North Rim. The next day crews at Bandelier National Monument faced the same passing cold front, lost their fire, and watched it burn into Los Alamos. Rick reckoned that burning would become much more difficult for the agency. He resigned and returned to Florida to conduct research burns for the Archbold Biological Station at Lake Placid. Within two years he moved to the Nature Conservancy, where he worked with Ron Myers and became fire management officer (FMO) for the southeast region, which also included the Caribbean and Central America, bringing him back to Belize. In 2005 he re-upped in the Park Service and what he calls his "briar patch" to become Everglades's fire ecologist. Then he took the job he was "born to do." In 2008 he became the park's FMO.

In reflective moments he muses about restoration and redemption and what faith and good works might do for a place that, as his footloose ancestors might have put it, was "rode hard and put away wet." As he sees it, in the early years the park burned too much. Then it burned too little. It's extraordinarily resilient in the face of fire, and not resilient in fire's absence. The agency should not seek to "restore" so much as sustain the "native" ecology—a deliberately vague goal—for which it must regain control over fire, not so much to reduce fire's damages as to exploit its benefits. Too often fire was seen by park administrators as an afterthought, as something that might control itself if the water regime got properly reestablished. Rick Anderson saw it as a positive force. Fire is what will jolt the reassembled parts to life. In the River of Grass fire must do what flooding does elsewhere: it flushes the channel clear of the debris that would otherwise dam it into senescence.

Yet he knows how often and easily the Everglades have frustrated the ideals, schemes, ambitions, and yearnings of those who have come to it to remake the land according to conceptions of agricultural reclamation, tourism, national parks, or wilderness. The national saga of fire unleashed, of fire suppressed, of fire tolerated, and of fire restored have each cycled through the Everglades, which received them and then rejected them in its time-honored way, by redirecting their consequences. It had defeated simple ideas of fire exclusion, of natural fire, of prescribed surrogates for restored fire, and might well do the same for ideas of redemptive fire. All were noble visions—ideals by which to enhance the Everglades, not chop it up or drain it off into sugarcane or cattle or commercial hunting. But the Glades do not distinguish among intentions. They only know what does or doesn't happen.

The outcome is uncertain. Too little is known and too much has been done to assert we can predict what will happen next, assuming it can even be done according to plan. But it seems wholly appropriate that someone of Rick Anderson's pedigree should oversee the next phase. Fire management in Florida has worked poorly when ideas get imposed from the outside; it has thrived best when old Florida hybridizes with newcomers or when indigenes cycle to the outside for an education before returning. Rick says he has no single vision. He wants to make fire serve the land, which means being nimble, being opportunistic, mixing fire types and timing, making up for what is lost in one place with gains in another. It means seizing control over a critical process for a landscape in rehab, or at least reasserting the significance of fire to the biota for whatever rehabilitation or restoration might come to it.

To most visitors the wild Everglades seem immutable, or they evolve with the sluggish tempo of its creeping waters. To those familiar with its biota, the place, as Rick puts it, "moves on fast forward." It moves with the speed of fire.

At the start of the new millennium two national endeavors were authorized that would affect restoration at Everglades. One was the Comprehensive

Everglades Restoration Plan, a 30-year federal-state undertaking intended to reinstate more of the historic hydrology. The other was the National Fire Plan (NFP), catalyzed into political existence by the horrific 2000 fire season. The NFP soon segued into the Healthy Forests Restoration Act of 2003; together they directed considerable attention and funding to fuels treatments, to presuppression projects to contain large wildfires, and, implicitly, into projects to return fire to fire-starved landscapes. That was also the year wildfires ripped unchecked through the Northern Rockies and an escaped prescribed fire blew through Los Alamos, and it was the year Rick Anderson showed you could come home again and returned to Florida.

Everyone could understand, or thought they understood, what restoring water meant. Ever since Marjory Stoneman Douglas's classic, the Everglades had been characterized as a River of Grass. (Douglas had written the book for the Rivers of America series, hence the title.) Rivers are about water. You restore a river by restoring its water. The idea that you might have to restore it by burning, that it needed fire to maintain its organic sediments at grade, was, at best, counterintuitive. It was axiomatic that burning could not save Everglades: rehydration had to come first. Rehydrating might even be sufficient since everything else would literally flow from it.

That was unlikely, and it was a misreading of the Everglades biota. The prevailing sense endured that fire mattered because it was a problem, or could become one; that fires had gone feral because the hydrology had failed; that prescribed fires were something you did because, if you didn't, even worse fires might result. There was little appreciation that fast fires complemented the slow waters. Moreover, restoring water did not threaten surrounding communities, endangered species, air quality, visitor safety, or staff. No one would, even in principle, wish its waters away from the Glades. But it was possible for many to imagine its fires gone, like the removal of an infestation of nasty exotics.

Yet the reality was that the Everglades was a dialectic of water and fire. Without fire the system would choke on its peat. Species would flee to find habitats that suited them. The saw grass—the very emblem of the Everglades—grew in ways that held oxygen in stem cavities such that it accelerated combustion. Saw grass grew to burn. Fire would not only come, as the rains would; it would overflow, as the Glades routinely did.

It was not simply inevitable: it was essential. Rehydration might hold the biotic pieces of the Everglades mosaic together, like a watery grout. But it would be fire that kindled them to life.

When Rick Anderson returned, then, the park was abuzz with plans for restoration. He thought in humbler terms—redemption rather than restoration. He knew that Everglades would not, even in principle, reflood with fire as it sought to do with water. There would be no $8.6 billion program to widen the borders for fire, to broaden the room for burning. The burning would fall behind. All-out restoration was a fool's errand. What made sense was to accept the broken land as it was, and as it might be after it was once again watered, and to think about what might be possible for renewal.

Renewal doesn't have the cultural cachet of restoration, particularly in the sense of rewilding. It doesn't come with the moral radiance surrounding redemption, and its aura of attrition and promise of atonement. But it is a way to keep fire on the land in ways that will allow the future to recover and to draft from flame in new and unanticipated ways. The land does far worse unburned than burned poorly. If it holds fire, it will remain malleable enough to be resilient and allow future fire practices to evolve. The goal of the draft 2010 fire plan would assure that fire and fire management capabilities were not lost.

That's not the fanfare of a slogan that will rally the public or rouse politicians. It's the hard-won wisdom of someone who grew up in the fading light of Old Florida and returned to a neon-lit New Florida. It's the distilled lore of a Floridian pedigree operating in an abused wetland that lacks the water to quench its ecological thirst, a vast oft-trashed marsh that fronts a ravenous megalopolis and two American Indian reservations full of toxic history, a landscape rife with threatened and endangered species, poised to be overrun with exotic pyrophytes and fauna, lusted after by developers of super jetports and the Turkey Point nuclear power plant that claims a right-of-way through Taylor Slough, and blessed and cursed by environmental groups with designations as a World Heritage Site, an International Biosphere Reserve, a Wetland of National Importance, and the Marjory Stoneman Douglas Wilderness. It's a singular place without analogues elsewhere in the world. It will require a fire program of equal singularity.

All in all, not a bad place for the flow of Florida fire history to end.

EPILOGUE

Florida Between Two Fires

FLORIDA IS AS GOOD a place as any to begin the story of an America passing between two fires. What challenges fire management elsewhere is present here, but what makes fire management work is also abundant, and that is particularly true for prescribed burning. If fire management fails in Florida, it will likely fail elsewhere. And if it succeeds, it may become a point of ignition to spread reform beyond the state's borders.

In many attributes Florida is as much an institutional outlier as the peninsula is a geographic appendage. It's now the fourth-most populous state and the second with the most threatened and endangered species (after Hawaii). Its economy, more mixed than many, is still biased toward services and the immigration of people and money, a financial system of transfer payments and remittances in reverse. For all its beaches and subtropical scrub, land use resembles the country at large, just warmer and more lush than most, and more at risk from the tantrums of water and fire. Sprawl is a byword, as elsewhere, as once-rural landscapes convert to super Walmarts, office parks, retirement communities, golf courses, theme parks, and other fantasies on swamps. Little of this colonization is in sync with the land since it relies on money made elsewhere. All this is a scenario familiar throughout the country, here exaggerated. But, unusually, it is countered in Florida by an aggressive program of nature protection not replicated by other states. Its enduring public themes are water and wildlife.

Florida's fire scene shares, or finds analogues for, these traits. It's a mixed economy, but skewed from the national norm by its promotion of prescribed burning. It ranks among the top in numbers of fires, first in acres burned by prescription, and first in what might be called fire density. It boasts one of perhaps four hearths nationally for an established fire culture. It counters the power of fire management on public lands by melding an emphasis on private landowners' right to burn. Over the past century it has repeatedly led the country in arguing for controlled fire as a basis for fire management; over the past few decades it has repeatedly transformed the exceptional into the exemplary and then exported that experience throughout the United States as well as the Caribbean and Mesoamerica. In the political economy of American fire, it marks one apex of the national fire triangle.

Yet for every fire prospect in Florida there is a fire problem, and together they alloy into what might be called Florida's fire paradox.

In 2008 a new generation of fire issues led worried leaders to organize a Fire Summit. Held at Tall Timbers, backlighted by a Great Recession that was, to general astonishment, stopping and even rolling back the long flood of migrants, the select group debated the "future of fire in Florida and Georgia." Whatever their historic feuds, however much their public agencies or private landownership differed on standards of fire management and what the future ought to look like, they all knew they shared a common cause in protecting prescribed fire for that future.[1]

The primary threats were two. One was familiar. It was, in fact, the basis for the state's economy, the influx of migrants and their conversion of a rural landscape informed by rural practices into urban centers or exurban enclaves. The newcomers saw the land differently, wanted to promote different values, but they lived on the land, not off it. They brought a heritage hostile or indifferent to fire. Their influx diced up landscapes, fragmenting them with roads and incombustible patches, eliminating the tolerance for spillage among neighboring burners, and removing the presence of timber companies and their firefighting capabilities from the scene. Left unanswered, these changes would yield more understory rough, more savage fires, and a reduced capacity to treat fuels and fight wildfires.

Fire managers would have to adapt their traditional solution, prescribed burning, to that contested fringe between suburb and woods—a tricky problem, but an avatar of what Florida fire managers had been doing for decades. The recession caused a temporary halt to the rush of development, although it also reduced the state's ability to respond.

The second threat was smoke, or more broadly, air quality. Unlike flames it was not possible to hold smoke at firelines; it would rise, drift, collect, saturate humid air masses with condensation nuclei, and literally blow with the wind. Both wildfires and prescribed fires had smoked in cities, sent asthmatics to emergency rooms, and facilitated car crashes on smoke-obscured highways. Its particulates crowded with particular force into the range of visible light; smoke could not, like automobile exhaust, merge invisibly into the general atmosphere. Its soot could cause ambient air to deteriorate beyond the tolerances permitted under the Clean Air Act. Moreover, emissions could be laden with toxins and carcinogens. They washed out quickly, but they were present. And fires released vast quantities of greenhouse gases. The fundamental chemistry was the same as hydrocarbons burned as coal; wood smoke was just more visible, which made it an easy target. Dale Wade and Hugh Mobley warned that, regardless of heritage or benefits to fuel management, if fire managers could not keep smoke out of sensitive areas, the public would shut burning down. Perversely, liability favored wildfire over controlled fire. Wildfire was an act of God or nature and thus granted automatic waivers. Prescribed fire identified a culpable agent.[2]

In order to address the novel threat from smoke, the Florida fire community reworked old arguments about fuels. The essence was prescribed smoke was better than wild smoke. The threats to public health would be less if the burning was domesticated than if it went feral. There were fewer injuries when Florida burned 2.7 million acres with prescribed fire in 2010 than when the wildfires of 1998 burned half a million acres. Instead of tightening the screws on prescribed fire, the Environmental Protection Agency (EPA) and local air-quality boards ought to be promoting more. And these arguments said nothing about the biological benefits of fire, for which there is no real alternative. The Fire Summiters concentrated on that volatile boundary where wildlands and houses met, which could no longer be a line drawn in the sand because the insouciant smoke crossed it with even more abandon than did flame. Fire's "interface" had become three dimensional. Florida's smoke entered a global airshed.

Crisis and response followed a familiar formula. Here, once more, was an attempt to impose a standard developed elsewhere onto a refractory Florida, an experiment that could only end with an unexpected and worsened outcome than intended. And, once again, the response was to fight first for an exemption, and then to turn the threat into an opportunity to devise arguments for prescribed fire and for approaches to smoke management that might be usable elsewhere. The EPA was invited to join the conversation, and was then absorbed, until a fusion resulted that allowed the FFS to act as a regulator of air quality. The new threats were thus redefined into variants of the old ones and could be handled the same way. Almost certainly the EPA will continue to evolve a modus operandi that will allow the burning to proceed, although with more restraints and the likelihood of an overall reduction in geographic range. Every house raised in the interior shrinks not only the scope for flame but for smoke.

And that is the secular trend that goes beyond the recycling of fire fads and problem avatars. Despite burning that exceeds state accomplishments anywhere else, both states declared their numbers inadequate. Georgia held 24.8 million acres of forested land and burned an annual average of 1.2 million acres. Florida had 21.7 million burnable acres and on average put over two million of them to the torch each year. Georgia wanted to double the amount burned by 2020. Florida wanted to shrink its current 10-year rotation for fire to three to five years, and in many places, one to two years would make for sounder ecology; this would require that it more than double its annual burning regime. Probably the public lands would succeed in boosting their burning. Private lands would burn minimally, or find alternatives in mechanical treatments or biofuels. For all their symbolic reach, the plantations of the Red Hills occupy fewer acres than Orlando. Unless managed for public purposes (such as Nature Conservancy sites), the overall prospects point toward more intensive fire management on the public estate and less on private.

The greater fissure may be intellectual: the failure to understand fire's macroecology. Even as automobiles compete with driptorches and fire's management boils upward in columns of smoke, the science of fire ecology continues to measure working landscapes against pristine ones and to assess effects as they appear on the surface. The deep dynamic behind fire's presence, however, is the contest between the combustion that powers the modern Florida economy, one based on fossil biomass, and the open burning that energized its former economy, one based on surface biomass. The question of how much fire belongs on the land will be determined by how much combustion goes on in power plants, cars, gas furnaces—the whole apparatus of power that determines how Floridians actually live.

When Earth's keystone species for fire decided to exercise its firepower in different ways, that choice has rippled through biosphere and atmosphere. It cascades freely through the political geography of fire in Florida today. Behind every challenge to open burning lies a changed reality in anthropogenic combustion habits. Yet fire ecologists persist in believing that the problem is that the public does not understand the natural role of fire. One could equally note that fire ecologists do not understand the actual dynamic of fire on Earth and instead deal with an idealized model of what they would like fire to be. The role of industrial combustion remains as invisible to them as that of most free-burning fire to the public.

More than any other state, save perhaps Louisiana, the competition between open burning and industrial combustion will decide the fate of Florida as a geophysical entity. Not the matches tossed from pickups but the internal combustion engines (ICE) running those trucks will govern whether Florida will burn or flood. The Pleistocene Ice Ages raised Florida from the ocean; a coming ICE age may submerge it. In this regard, as in so many other themes in fire's presence, Florida remains not only a bellwether but an illuminating exaggeration.

In truth, for all its enduring presence on the Florida landscape, fire remains remarkably invisible to most of those people who live there. It appears during episodes of mass conflagrations as in 1998, a kind of fiery counterpoint to hurricanes; and it appears where modern Florida rubs against old Florida and prescribed burning seeks to broker between them. But even as I drove the great arc from Pensacola to Miami, fire faded into woods and cloud.

The burning that was going on took place mostly outside the public view—that was part of the prescription. Practitioners sought to limit the size of the patches burned and to disperse their smoke. Here and there, like summer thunderheads, smokes popped up along the south Florida horizon, but they soon blurred into the clouds and murky sky. Unless you looked, you would likely not spy them. In the Southwest the dry air and long vistas make every smoke stand out like a mountain silhouette. In Florida they are a thickening of the normal atmosphere, as flames are of the biota. The paradox is that keeping fire out of public sight leaves it out of the public mind.

In 1938, three years after the Big Scrub fire set national records for ferocity, Marjorie Kinnan Rawlings published *The Yearling*. An instant classic and national bestseller, the novel's plot hinged on a year in the life of the Baxters, a backwoods Floridian family, a detailed almanac of Florida natural history so rich it makes Thoreau's journals look like distracted doodlings. The story records the flowering and fruiting of sweet magnolia and chinaberry. It incorporates the seasonal cycle of storms and births and hibernation. It depicts floods, withered vines, and wildlife epidemics. Every delight has its horror: for each raccoon kit or fawn, there are rogue bears and rattlesnakes. The Baxter farm is attacked by wolf packs, foxes, and Old Slewfoot, the shrewdest black bear in Florida's literary imagination. But nowhere does fire appear on the land as either native resident, farm tool, or alien terror. The smoke of the Big Scrub burn was probably visible from Cross Creek, where and when Rawlings was writing, but neither it nor its domesticated kin burned into the text. Flame appears only once, when the Forrester boys in a drunken frolic maliciously burn down a house.

So, too, fire remains invisible in Florida history and consciousness. The achievement of Florida's fire communities has been to make it sufficiently visible to permit it to continue and to make Florida's management of fire visible to the nation.

The political geography of Florida has long assumed an exaggerated significance for the country overall. When it was a colony being swapped between Britain and Spain, worried expansionists likened it to a "pistol pointed at the heart of the United States." The panhandle barrel pointed directly at New Orleans and could thus threaten trade throughout the entire watershed of the Mississippi River. Whatever else its merits (or liabilities), the place was too important to overlook.[3]

Today, Florida affects the national scene as a hinge state, vital to the workings of Congress and indispensable to any successful presidential campaign. To play again on its shape, it has become the hanging chad of American politics. And so it is with fire policy. However confusing the issues, no other state has so forcefully put the matter of prescribed fire before the electorate. How the votes get tallied will say much about the future direction of fire in America.

NOTE ON SOURCES

MORE THAN ANY PREVIOUS BOOK I've written, my sources have been people. I acknowledge their generosity in the particular essays to which they contributed, but I need to extend a special thanks to Jim Brenner for helping make many of those contacts and for organizing a mini-gathering of fire folks in Tallahassee. The act is in keeping with his many efforts over many many years to advance fire management in Florida.

I've supplemented those conversations across the state with a few general references and a lot of specialized studies, much of it gray literature. I found the most informative of the general sources to be Ronald L. Myers and John J. Ewel, eds., *Ecosystems of Florida* (Orlando: University of Central Florida Press, 1990); Edward A. Fernald and Elizabeth D. Purdum, eds., *Atlas of Florida*, rev. ed. (Gainesville: University Press of Florida, 1996); David Colburn and Lance deHaven-Smith, *Florida's Megatrends*, 2nd ed. (Gainesville: University Press of Florida, 2010); and Baynard Kendrick and Barry Walsh, *A History of Florida Forests* (Gainesville: University Press of Florida, 2007).

Specific references are included with each essay. I find the gray literature particularly helpful, and these are kept—when they are kept at all over time—in files on site.

NOTES

PROLOGUE

1. The best summary of the relevant natural history is Ronald L. Myers and John J. Ewel, eds., *Ecosystems of Florida* (Orlando: University of Central Florida Press, 1990).
2. David Colburn and Lance deHaven-Smith, *Florida's Megatrends*, 2nd ed. (Gainesville: University Press of Florida, 2010), 15.
3. Excellent for developments since the 1970s is Jim Brenner and Dale Wade, "Florida's Revised Prescribed Fire Law: Protection for Responsible Burners," in *Proceedings of Fire Conference 2000: The First National Congress on Fire Ecology, Prevention, and Management*, ed. K. E. M. Galley et al. (Tallahassee, FL: Tall Timbers Research Station, 2003), 132–36.

AFTER THE REVOLUTION

1. My thanks to Ron Masters and Lane Green, who answered my questions with patience and precision, and to Juanita Whiddon, who generously made available the station's archives. Additional thanks to Ron, Lane, and Bill Palmer for comments on an earlier draft.
2. The actual source is uncertain, having become folklore itself. See "Frequently Asked Questions About Mead/Bateson," The Institute for Intercultural Studies, accessed August 23, 2015, http://www.interculturalstudies.org/faq.html.

3. The best summary remains Edward V. Komarek Sr., *A Quest for Ecological Understanding: The Secretary's Review, March 15, 1958–June 30, 1975* (Tallahassee, FL: Tall Timbers Research Station, 1977). For Stoddard's story in his own words, see Herbert L. Stoddard, *Memoirs of a Naturalist* (Norman: University of Oklahoma Press, 1969).
4. Bethea's quote appears in Baynard Kendrick and Barry Walsh, *A History of Florida Forests* (Gainesville: University Press of Florida, 2007), 431.

INTO THE OPEN AIR

1. I would like to thank Michael Dueitt, Jim Ferguson, Greg Seamon, Jim Durrwachter, and Miranda Stuart for taking time away from their preparations for a new course to talk with an old smokechaser who had turned to the dark side and become an academic. I should have been in the field.
2. C. Northcote Parkinson, *Parkinson's Law and Other Studies in Administration* (Boston, MA: Houghton Mifflin, 1957; repr., Cutchogue, NY: Buccaneer Books, 1996). The relevant essays are "Parkinson's Law, or the Rising Pyramid," 2–13, and "Plans and Plants, or The Administration Block," 59–69. Citations refer to the Buccaneer edition.
3. My survey relies on conversations with current staff and especially PFTC, "Prescribed Fire Training Center: 1998 thru 2008 Cumulative Summary," supplemented by reports for the years 2009 and 2010.

OUR PAPPIES ARE STILL BURNING THE WOODS

1. John P. Shea, "Our Pappies Burned the Woods," *American Forests*, April 1940, 159–62, 174.
2. For background information on this project, see Stephen J. Pyne, *Fire in America* (Princeton, NJ: Princeton University Press, 1982), 171–72.
3. Stoddard, *Memoirs of a Naturalist*, 245.
4. See Eric Hobsbawm and Terence Ranger, eds., *The Invention of Tradition* (Cambridge: Cambridge University Press, 1992).

NOT EVEN PAST

1. I would like to thank the staff of the Joseph W. Jones Ecological Research Center for their invitation to visit and for their introduction into the integrated history of Ichauway Plantation. Special thanks to Lindsay Boring, Mark Melvin, Steve Jack, Kay Kirkman, Jonathan

Stober, Scot Smith, and Jimmy Atkinson for their intensive tutorial into the character of fire and its management on the site. Any errors of fact or interpretation are of course mine alone.

2. On the development of the plot, see R. J. Mitchell et al., "Silviculture That Sustains: The Nexus Between Silviculture, Frequent Prescribed Fire, and Conservation of Biodiversity in Longleaf Pine Forests of the Southeastern United States," *Canadian Journal of Forest Research* 36 (2006): 2724–36; L. Katherine Kirkman et al., "The Perpetual Forest: Using Undesirable Species to Bridge Restoration," *Journal of Applied Ecology* 44 (2007): 604–14; R. J. Mitchell, L. K. Kirkman, and S. D. Pecot, "Restoration of Multi-Aged Longleaf Pine Woodlands Through Time" (unpublished study, Jones Ecological Research Center, n.d.).
3. An interesting sketch of the growth of quail plantations, and the Cooperative Quail Study Investigation, is available in Komarek, *Quest for Ecological Understanding*, 13–22. For a thumbnail of the Grouse Commission, see Stephen J. Pyne, *Vestal Fire* (Seattle: University of Washington Press, 1997), or for the source, Great Britain Board of Agriculture and Fisheries, Committee on Inquiry on Grouse Disease, *The Grouse in Health and in Disease* (London: Smith, Elder, 1911).
4. For a synopsis of longleaf as a conservation priority, see L. Katherine Kirkman, "Conserving Biodiversity in the Longleaf Pine Savannas of the USA," *Plant Talk* 39 (January 2005): 30–32. For a more thorough but popular history, see Lawrence S. Earley, *Looking for Longleaf: The Fall and Rise of an American Forest* (Chapel Hill: University of North Carolina Press, 2004).
5. J. Stober, "The Prescribed Fire Management Program at Ichauway 1994–2008" (Newton, GA: Jones Ecological Research Center, March 2009).
6. A good summary of Stoddard's role is Alfred G. Way, "Burned to be Wild: Herbert Stoddard and the Roots of Ecological Conservation in the Southern Longleaf Pine Forest," *Environmental History* 11, no. 3 (July 2006): 500–526. For the source, see Stoddard, *Memoirs of a Naturalist*.
7. See R. K. McIntyre et al., *Multiple Value Management: The Stoddard-Neel Approach to Ecological Forestry in Longleaf Pine Grasslands* (Newton, GA: Jones Ecological Research Center, June 2008).

THE FLORIDA FOREST SERVICE

1. A special thanks to Jim Brenner who helped organize my entire trip and showed bottomless patience in educating an outsider into the

hammock-like complexities of Florida fire. John Saddler helped, particularly with statistics, and Dennis Hardin contributed a valuable perspective on burning on state lands.

2. Major sources consulted include Kendrick and Walsh, *History of Florida Forests*; E. Dennis Hardin, "Institutional History of Prescribed Fire in the Florida Division of Forestry: Lessons from the Past, Directions for the Future" (unpublished essay); Ramona McClennan, *History of Fire Control, State of Florida* (n.p.: Florida Division of Forestry, Fire Control Bureau, February 12, 1973); C. H. Coulter, *History of the Florida Forest Service* (n.p.: Florida Forest Service, September 4, 1958); and R. A. Bonninghausen, "The Florida Forest Service and Controlled Burning," in *Proceedings, First Annual Tall Timbers Fire Ecology Conference* (Tallahassee, FL: Tall Timbers Research Station, 1962), 43–65.
3. On the origins of prescribed fire within the FFS, see John Bethea, quoted in Kendrick and Walsh, *History of Florida Forests*, 433.
4. Jim Brenner, conversations with author, January 2011.
5. The quote is from an anonymous reader solicited by the University of Arizona Press. It summarized so succinctly the process that I decided to simply quote. My thanks—doubly—to whoever wrote the comment.
6. In January 9, 2008, smoke from a prescribed fire set by the Florida Fish and Wildlife Conservation Commission, which then escaped, contributed to a superfog effect that ended with a 75-car pileup on I-4 and five deaths. The incident is a reminder that the Florida Prescribed Fire Act has not been tested in courts.

INTERLUDE

1. Raymond M. Conarro, "The Place of Fire in Southern Forestry," *Journal of Forestry* 40, no. 2 (February 1942): 129–31.
2. Ibid., 6.

THE MORE THINGS CHANGE

1. Special thanks to Kevin Hiers and Brett Williams, and the rest of the Eglin fire crew, for a study tour of the records at Jackson Guard and a bracing day of burning.
2. Inman F. Eldredge, quoted in Elwood R. Maunder, *Voices from the South: Recollections of Four Foresters* (Santa Cruz, CA: Forest History Society, 1977), 35. The interview, 48 years later, complements Eldredge's

published classic, "Fire Problem on the Florida National Forest," *Proceedings of the Society of American Foresters* (Washington, DC: Society of American Foresters, 1911), 164–71.

3. Eldredge, "Fire Problem," 167–68.
4. Ibid., 168.
5. Ibid., 167.
6. An excellent chronicle of human history and forest management is available in Appendix A, Eglin Air Force Base, FL, "Integrated Natural Resources Transitional Plan, 1998–2001."
7. The announced purposes of Eglin keep evolving. The quote comes from an account of "installation history" available among the historical files available at Jackson Guard. An excellent summary of uses is "Integrated Natural Resources Management Plan: Eglin Air Force Base, FL, 2002–2006." Endorsed plan to satisfy requirements of Sikes Act. Eglin Air Force Base, 2002.
8. Sikes Act of 1960, 16 U.S.C. § 670 (1960).
9. Louis Provencher et al. 2001. Restoration of Fire-Suppressed Longleaf Pine Sandhills at Eglin Air Force Base, Florida. Final Report to the Natural Resources Management Division, Eglin Air Force Base, Niceville, Florida. Science Division, The Nature Conservancy, Gainesville, Florida.

REGIME CHANGE

1. Greg Titus, Pete Kubiak, Joe White, Doug Scott, Robin Will, and Joe Reiman made my understanding of fire at St. Marks possible. The refuge fire crew provided a Greek chorus to our declamations. My thanks to them all. Any misunderstandings are of course entirely my own responsibility.
2. John J. Lynch, "The Place of Burning in Management of the Gulf Coast Wildlife Refuges," *Journal of Wildlife Management* 5, no. 4 (October 1941): 454–57. Notes on early days from discussions with former practitioners at St. Marks on January 5, 2010.
3. Joe White, "On Frank's Shoulders," *The Eagle Eye* (newsletter, St. Marks National Wildlife Refuge, Winter 2010–11), 10–11.
4. For an intriguing survey of the concept's origins, see Patrik Krebs et al., "Fire Regime: History and Definition of a Key Concept in Disturbance Ecology," *Theory in Biosciences* 129, no. 1 (2010): 53–69.

   My own interest came from grad courses in fluvial geomorphology, in which I learned how French hydrologists characterized rivers as "*en régime*" and of course from history, where it was common to speak of ancien régimes. It was easy to make the analogic leap from such expressions to fire. I then discovered Gill's papers.

5. See M. L. Heinselman, "Fire Intensity and Frequency as Factors in the Distribution and Structure of Northern Ecosystems," in *Proceedings of the Conference, Fire Regimes and Ecosystem Dynamics*, ed. H. A. Mooney et al. (Washington, DC: U.S. Forest Service, 1981), 7–57, and K. M. Davis and R. W. Mutch, "Applying Ecological Principles to Manage Wildland Fire," *M580: Fire in Ecosystem Management* (training course, National Advanced Fire and Resource Institute, Tucson, AZ, 1994).

EAST IS EAST, WEST IS WEST

1. Thanks to Brigham Mason, ranch wildlife biologist, for an informative introduction to ranch fire practices and to Sisters Miller and Zollinger, for an introduction to ranch history.
2. Frederic Remington, "Cracker Cowboys of Florida," *Harper's* 91, no. 543 (August 1895): 339–45.
3. Two magazine articles usefully examine the ranch: Cynthia Barnett, "The Church's Ranch," *Florida Trend*, December 2001, 56–61, and Kyle Partain, "Cattle Kingdom," *Western Horseman* 73, no. 6 (June 2008): 34–40.
4. See, for example, Karen M. Bradshaw, "Backfired! Distorted Incentives in Wildfire Suppression Techniques," *Utah Environmental Law Review* 31, no. 1 (2011): 155–79.

A TALE OF TWO LANDSCAPES

1. J. C. Ives, *Memoir to Accompany a Military Map of the Peninsula of Florida, South of Tampa Bay* (New York: Wynkoop, 1856). The quotations are from page 1.
2. This essay is possible because of the generosity of Robert Dye, who shared a day in the field with me, and then—unusually, for my tour of Florida—handed over documentation. The various publications of Paula Benshoff also proved vital in appreciating what it takes to move ideas into practice. The history of fire in the region presented in this essay is basically theirs, recast into a somewhat different frame, one that I believe they do not wholly agree with. They deserve credit for what works in the essay. For what doesn't, the fault is mine.

3. For a description of the Myakka Island concept, see P. J. Benshoff, *Myakka*, 2nd ed. (Sarasota, FL: Pineapple Press, 2008).
4. Quoted in David Nelson, "The Great Suppression: State Fire Policy in Florida, 1920–1970," *Gulf South Historical Review* 21 (2006): 89.
5. James A. Stevenson, "Conference Dedication of E. V. Komarek, Sr.," in *Proceedings, 17th Tall Timbers Fire Ecology Conference: High Intensity Fire in Wildlands: Management Challenges and Options* (Tallahassee, FL: Tall Timbers Research Station, 1991), 1.
6. The Buck Mann story is from Benshoff, *Myakka*, 191.
7. The burning pattern is from Mike Kemmerer, e-mail message to Robert Dye, January 12, 2011.
8. For a good digest of the fire scene as the principal agents of the Myakka fire program saw it, see P. J. Benshoff, R. Dye, and B. Perry, "Fire in the Florida Landscape" (Myakka River State Park, n.d.). For an overview of the ecology, see *Proceedings of the Florida Dry Prairie Conference: Land of Fire and Water: The Florida Dry Prairie Ecosystem* (DeLeon Springs, FL: Painter, 2004), http://www.ces.fau.edu/fdpc/proceedings.php.
9. Quote from Robert Dye, e-mail message to author, March 7, 2011.
10. For a detailed rendering of some of the experiments, see Robert Dye, "Use of Firing Techniques to Achieve Naturalness in Florida Parks," in *Proceedings, 17th Tall Timbers Fire Ecology Conference: High Intensity Fire in Wildlands: Management Challenges and Options* (Tallahassee, FL: Tall Timbers Research Station, 1991), 353–60.
11. Numbers from Florida Park Service, *Resource Management Annual Report: July 2009–June 2010*. (Tallahassee, FL: Bureau of Natural and Cultural Resources, [2010?]).
12. Stevenson, "Conference Dedication," 2.

ONE FOOT IN THE BLACK

1. Paula Seamon, Zach Prusak, Walt Thomson, and Ron Myers made this essay possible, and supplied documentation to back up their remembrances. I'm grateful to them all, not least for their patient tolerance toward my efforts to mash four essays into one. TNC, the Florida chapter, the fire initiatives, Ron—each deserves its own panegyric. Which I'll try to do, all in good time.
2. Mark Heitlinger with assistance from Allen Steuter and Jane Prohaska, "Fire Management Manual," revised September 1985 (The Nature Conservancy).

3. See TNC, "Controlled Burning: Getting It Done" (TNC, 2009) and "A Decade of Dedicated Fire: Lake Wales Ridge Prescribed Fire Team" (TNC, 2010); "The Northeast Florida Resource Management Partnership, Draft Final Report (April 1, 2008–December 31, 2010)"; "Central Florida Ecosystem Support Team, Florida Fish and Wildlife Conservation Commission, Final Report (February 1, 2010–December 31, 2010)" (January 2011).
4. Two publications describe the program: Ayn Shlisky et al., *Fire, Ecosystems and People: A Preliminary Assessment of Fire as a Global Conservation Issue* (TNC, 2004) and Ronald L. Myers, *Living with Fire—Sustaining Ecosystems and Livelihoods Through Integrated Fire Management* (TNC, 2006).
5. See Todd Wilkinson, "Prometheus Unbound," *Nature Conservancy*, May/June 2001, 12–20, for a survey of the national scene at the time, and in the same issue, William Stolzenburg, "Fire in the Rain Forest," 22–27, for its work in Mesoamerica.

FIRE 101 AT STAR FLEET ACADEMY

1. Thanks to Michael Good, Fred Adrian, and Dorn Whitmore for their patient (and occasionally colorful) explication of fire management at Merritt Island, and to Robert Eaton for editorial comments.
2. On the fatal fire, see "Report of Board of Inquiry on Merritt Island NWR Wildfire Fatalities," Memorandum from Regional Director, FWS, Atlanta, Georgia, to Director, FWS, Washington, DC (July 10, 1981) and Sebastino J. Castro and C. R. Anderson, *A Report of the Committee on Appropriations, U.S. House of Representatives, on Wildfire on Merritt Island*, Surveys and Investigations Staff, House Appropriations Committee, December 1981.

UNDER THE DOME

1. Thanks to Adam Watts, Pedro Ramos, John Nobles, Mindy Wright, Jim Snyder, and Justin Turnbo for exceptional patience in describing what they do and what they have learned. It reminded me how much I miss fire shoptalk and listening to the rhythms and slang and blended stories of people doing fire on the ground. Our conversation became a porous border to my own past.

2. See James A. Farr and O. Greg Brock, "Florida's Landmark Programs for Conservation and Recreation Land Acquisition," *Sustain* 14 (Spring/Summer 2006): 36–45. The essay summarizes the chronology of acquisition programs and provides the quoted numbers.
3. The best summary of Big Cypress, both natural and human history, remains Michael J. Duever et al., *The Big Cypress National Preserve*, Research Report No. 8 (National Audubon Society, 1979; second printing, 1985); the document includes a chapter on fire by Ronald L. Myers. The Preserve's "Fire Management Plan 2010" updates and distills this material and gives it an operational focus.

THE EVERBURNS

1. Marjory Stoneman Douglas, *The Everglades: River of Grass* (Port Salerno: Florida Classics Library, 2002), 5.

   Special thanks to Rick Anderson for a delightful day of meditations, field exhibitions, and stories. Thanks, too, to Jennifer Adams, Bonnie Ciolino, and other members of the fire management staff for help in tracking down documents and getting them copied. Anyone familiar with the Everglades fire scene will recognize my reliance on Dale L. Taylor, *Fire History and Fire Records for Everglades National Park 1948–1979*, Report T-619, South Florida Research Center, Everglades National Park (1981), which summarizes the known records. Beyond that I have referred to the 1991 and 2010 (draft) fire management plans, which summarize much of the previous history. There isn't much in park records that they miss. The best distillation of the basic fire ecology remains Dale Wade, John Ewel, and Ronald Hofstetter, *Fire in South Florida Ecosystems*, U.S. Forest Service General Technical Report SE-17 (Asheville, NC: Southeast Forest Experiment Station, 1980). The park filing cabinets are veritable hammocks of peaty grey literature.
2. Quotes from Daniel B. Beard, "Let 'er Burn?," *Everglades Natural History* 2, no. 1 (March 1954): 2–8; Guy J. Bender, "The Everglades Fire Control District," 1941, copy in the files of the Everglades Fire Management Office; Daniel B. Beard, *Wildlife Reconnaissance: Everglades National Park Project*, Department of the Interior, National Park Service, October 1938, 51–52.
3. Douglas, *Everglades*, 349, 376, 376.
4. Beard, "Let 'er Burn?," 8; Beard, *Wildlife Reconnaissance*, 51.

5. William B. Robertson Jr., "A Survey of the Effects of Fire in Everglades National Park," unpublished report to the National Park Service, submitted February 5, 1953.

EPILOGUE

1. Joe Michaels, Gail Michaels, and Ron Masters, eds., *Fire Summit: The Future of Fire in Florida and Georgia, January 16–18, 2008* (Tallahassee, FL: Tall Timbers Research Station, 2008).
2. Dale Wade and Hugh Mobley, *Managing Smoke at the Wildland-Urban Interface*, General Technical Report SRS-103, U.S. Forest Service, 2007, i.
3. The pistol allusion was general in the early 19th century, quoted in William H. Goetzmann, *When the Eagle Screamed: The Romantic Horizon in American Diplomacy* (1966; repr., Norman: University of Oklahoma Press, 2000), 4.

INDEX

ABOUT THE AUTHOR

Stephen J. Pyne is a professor in the School of Life Sciences, Arizona State University. He is the author of over 25 books, mostly on wildland fire and its history but also dealing with the history of places and exploration, including *The Ice*, *How the Canyon Became Grand*, and *Voyager*. His current effort is directed at a multivolume survey of the American fire scene—*Between Two Fires: A Fire History of Contemporary America*, and *To the Last Smoke*, a suite of regional reconnaissances, all published by the University of Arizona Press.